Library Research Guide to Biology

"Library Research Guides" Series

JAMES R. KENNEDY, JR. and
THOMAS G. KIRK, JR., Editors

No. 1 LIBRARY RESEARCH GUIDE TO RELIGION AND THEOLOGY

No. 2 LIBRARY RESEARCH GUIDE TO BIOLOGY

No. 3 LIBRARY RESEARCH GUIDE TO EDUCATION

Library Research Guide to Biology

Illustrated Search Strategy and Sources

by
THOMAS G. KIRK, JR.
Science Librarian
Earlham College

("Library Research Guides" Series, No. 2)

Pierian Press
ANN ARBOR, MICHIGAN

Published August 1978.

International Standard Book Numbers: 0-87650-098-X (cloth);
0-87650-099-8 (paper)
Library of Congress Catalog Card Number: 78-61710

Pierian Press, P.O. Box 1808, Ann Arbor, Michigan 48106
Printed in the United States of America

Contents

Preface

For Teacher and Librarian

Library instruction includes much more than the teaching and learning of specific facts about reference books, the card catalog, and other library tools. Beyond these basics, instruction should include a concept of the organization of libraries and the classes of literature and their relationships. This additional knowledge is necessary if students are to do more than just browse through a few books or use the card catalog casually. Users of the library should have familiarity with the procedures used to find information: where to start, how to proceed from any given point, and what type of source will give the information needed. The attainment of this kind of knowledge is based on the concept that the library is ". . . a highly complicated system, or better, a network of inter-related systems, which organizes and controls all kinds of communication."[1]

The techniques used in an effective library search are urgently needed by a student today and in his role of citizen tomorrow. The trend toward increased use of independent study as one method of acquiring an undergraduate education, means that students have an immediate need to learn to use the library.[2] In addition, college faculty have come to realize that the accumulation of facts is not enough. Increasingly, attention is being paid to interrelated ideas, principles and generalizations, and to ways of looking at information.[3] These two trends mean that students have an immediate need to know how to use the library. Libraries interested in serving their clientele must recognize and fulfill this need.

While this immediate need is pressing upon the student, he and librarians may not be as aware of the recognized need for continuing education.[4] In earlier years the idea of continuing adult education was apt to be an educator's idea of the utopia with its educated citizenry. Today, in a real sense, that utopia has become a reality. It is quite clear that the growth in the amount of printed material[5] means that it is no longer possible to learn all a person will ever need to know during one stay in college early in his or her life. Therefore continuing education has increasingly become a necessity. From what has been said above, it is absolutely clear that there are compelling reasons for library instruction to undergraduates in college and universities today. These compelling reasons include the trend toward greater use of independent study, the de-emphasis of facts and emphasis on principle, and the greater number of persons involved in continuing education in later adult life.

The librarian must spend a great deal of his or her time preparing and giving library instruction if the student is to achieve the desired sophistication in his or her library use. Unfortunately, this would mean greatly increased staff for most libraries and therefore additional expenses which most institutions cannot afford. It is therefore necessary to make available some sorts of teaching devices that can be used in place of the extensive hours that would be required of the librarian or teacher. This research guide is an attempt to provide such a teaching device.

Students will make the most effective use of this guide if it is a required assignment and is used in conjunction with another assignment which requires use of the library. The faculty and librarian must convey the idea that all library assignments are an important component of the course. In addition, it must be made clear to the students that effective use of the library is one of the objectives of the course.

Because most libraries are organized alike but have minor differences, it might be concluded that it is easy to design a guide that can be widely used. The truth is, however, that the "minor" differences are not so minor. However, a guide which is too general would lack certain essential features, the most important of which is the imitation of an actual search. Thus, while the guide has been carefully written to be as general as is practically possible, it will nevertheless be necessary for a particular library to have certain basic features. These are listed and described below. The alternative is to modify the guide to suit particular libraries. This might be done by replacing certain pages with locally written ones.

1. *Card Catalog*. It will be necessary for the library to have a card catlog. Whether it is divided or dictionary makes no difference. It is equally important that Library of Congress subject headings are used and that a copy of the list of headings be available at the public catalog (*Library of Congress Subject Headings*, 8th or later ed., 1975, Washington, D.C. Library of Congress). If a modified list is used, then a locally produced list should be made available to the catalog users. Each card in the catalog should have tracings on it. Therefore, it will be possible to identify all subject headings under which a particular book is listed regardless of the entry under which it was first located.

2. The Library should have the *Science Citation Index*, if Chapter 5 is to be useful to the student.

3. The library should have *Biological Abstracts, Biological and Agricultural Index. Biology Digest*, and/or the

"Permuterm Index" of *Science Citation* Index if Chapter 6 is to be useful to the student.

4. The library should have copies of
   a) *McGraw-Hill Encyclopedia of Science and Technology*. 3rd or 4th ed. 15 vol. New York: McGraw-Hill, 1971 or 1977.
   b) Gray, Peter, *Encyclopedia of the Biological Sciences*. 2nd ed. New York: Van Nostrand Reinhold, 1970.

In Appendix I there is a test on the catalog card and *Readers' Guide* on which someone familiar with these two library tools should do well. It is recommended that users take the test before starting to read the guide. If you are using the guide in a formal course setting, you may want to use this test, or one of your own construction, to decide whether some of your intended audience need preliminary instruction on the basic mechanics of library use.

This *Library Research Guide* is the result of twelve years experience in teaching undergraduate science students – both majors and non-majors – the intricacies of the scientific literature and getting access to it. I believe given motivation and stimulation to use the library, this *Guide* can be an effective teaching instrument. Your experiences with the *Guide*, errors you detect, and comments on teaching patrons to be effective library users – will be gratefully received.

Thomas G. Kirk

1. Knapp, P.B., *The Monteith College Library Experiment*, p.40, (Citing Progress Report, no. 2, pp. 19-20, by Patricia B. Knapp).
2. Snyder, Luella, *The Second Kind of Knowledge*, p.3.
3. Knapp, P.B., "The Library's Response to Innovation in Higher Education," *California Librarian* 29:147, April, 1968.
4. Bell, Daniel, "Reforming General Education," in *Improving College Teaching*, p. 351.
5. The number of new titles and editions published in the United States today is up 36% from 30,050 in 1968 to 40, 846 in 1975. (Figures are from *Bowker Annual*, 1968, p. 61 and 1975, p. 179.) The amount of scientific literature is doubling every 10-15 years. (From *McGraw-Hill Encyclopedia of Science and Technology*, 4th ed., v.7, p. 612.)

Preface

For Students

The research guide contained here is intended for the student of biology who, for the first time, is approaching a problem that requires sophisticated use of the library. The guide assumes that the student has had previous experience in using the library for assignments requiring some use of the card catalog, and the *Readers' Guide to Periodical Literature*. In order to give the student an opportunity to evaluate his or her competence, a library test covering the card catalog and *Readers' Guide* is included in this volume as Appendix I. Immediately following the test are the correct answers. The experience of librarians who have used the test would suggest that a person scoring below 8 should have some basic instruction in the use of the library before using this guide. If you score below 8 on the test, you should see a reference librarian in your library or consult Margaret G. Cook, *The New Library Key*, 3rd ed. (New York: Wilson, 1925), or Ella V. Aldrich, *Using Books and Libraries*, 5th ed. (Englewood Cliffs, New Jersey: Prentice-Hall, 1967).

After you have successfully completed the library test you are ready to use this guide.

If you need to know the general procedures for writing term papers, including notetaking, outlining, and bibliographic form, use this book in conjunction with: Kate L. Turabian, *Students' Guide for Writing College Papers*, 2nd ed. (Chicago: University of Chicago Press, 1969); or Lucille Hook and Mary V. Garver, *The Research Paper*, 4th ed. (Englewood Cliffs, New Jersey: Prentice-Hall, 1969).

As a biology student you will have some special problems in writing a scientific paper for which these books will not be helpful. If you need information on style, organization and referencing procedures in biology papers, see the Council of Biology Editors, Committee on Form and Style, *CBE Style Manual*, 3rd ed. (Washington: American Institute of Biological Sciences, 1972).

Acknowledgments

The preparation of a book, even one as modest as this, is the result of many hours of work. Not just by the author, but by many behind--the--scenes people. To Vickie Storck, Eira Longstreet, Carol Lebold, Anne Hudson and Louise Hinkley has gone the task of typing the many versions of the manuscript; and to Lynda Curry the difficult chore of cleaning up someone else's writing style -- mine. To each of them I am indebted.

James Kennedy was immensely helpful by "breaking ground" for the Series which has made my job considerably easier. From his experienced perspective he was a source of sound and thoughtful advice.

The staff in the Science Division at Miami University's King Library were most helpful by extending favors to me while I used their collection in writing this book.

Finally thanks go to Evan Farber and the learning community at Earlham College who encourage one to use all of his creative talents and energies for the sake of young peoples' education. And in return I am continually educated by my efforts.

Credits for Figures

Thanks are also due to the many publishers cited below who gave their permission to use excerpts from copyrighted works. Without their courtesy this book could not have been the illustrated guide that was intended. Uncopyrighted materials are also cited below in order to make the list of figures complete.

Unnumbered figure: B.C. by permission of John Hart and Field Enterprises, Inc.

Figures 1-4: From *McGraw-Hill Encyclopedia of Science and Technology*. 15 vols. 4th ed., Copyright 1977. Used with permission of McGraw--Hill Book Co.

Figures 5-6: From *Encyclopedia of the Biological Sciences*, 2nd ed. by Peter Gray. Copyright 1970 Litton Educational Publishing, Inc. Reprinted by permission of Van Nostrand Reinhold Company.

Figures 7-10: From *Animal Life Encyclopedia*, 1972-1976, vol. 13 by D. Bernhard Grzimek. Copyright 1972 Litton World Trade Corp., 1968 Kindler Verlag A.G. Zurich. Reprinted by permission of Van Nostrand Reinhold Company.

Figures 11-12: United States Library of Congress catalog cards.

Figure 13: B.C. by permission of Johnny Hart and Field Enterprises, Inc.

Figure 14: From United States Library of Congress, *Library of Congress Subject Headings*. 8th ed., 1975.

Figure 15: United States Library of Congress catalog cards.

Figures 16-18: From *Ecology and Field Biology* by Robert L. Smith. Copyright 1966 by Robert L. Smith. Used with permission of Robert L. Smith.

Figures 19-21: From *Fundamentals of Ecology*, 3rd ed. by Eugene P. Odum. Copyright 1971 by W.B. Saunders Company. Used with permission of W.B. Saunders Company and Eugene P. Odum.

Figures 22-23: From *Wildlife Ecology: an Analytical Approach*, by Aaron N. Moen. W.H. Freeman and Company. Copyright 1973. Used with permission of W.H. Freeman Company.

Figure 24: Summary outline, key terms and references.

Figures 25-28: From *Advances in Ecological Research*. vol. 9. Copyright 1975 by Academic Press Inc. (London) Ltd. Used with permission of Academic Press Inc. (London) Ltd.

Figure 29: Summary outline, key terms, and references (revised).

Figure 30: How a citation is indexed in *Science Citation Index*. A portion from *Ecology*, vol. 57. Copyright 1976 by Duke University Press. Used with permission of Duke University Press.

Figures 31-34: From *Science Citation Index*, 1972, 1974. Copyright 1972, 1974 Institute for Scientific Information. Used with permission of the Institute for Scientific Information.

Figures 35-44: From *Biological Abstracts*, vols. 57, 59 and 60. Copyright 1974, 1975 by Biosciences Information Service of Biological Abstracts. Used with permission of BIOSIS.

Figure 45: From *Bioresearch Index*, vol. 12. Copyright 1976 by Biosciences Information Service of Biological Abstracts. Used with permission of BIOSIS.

Figures 46--47: From *Biological and Agricultural Index*, copyright 1972, 1973, 1974, 1975. Material reproduced by permission of the H.W. Wilson Company, publisher.

Figures 48-49: From *Biology Digest*, vol. 1, copyright 1974, 1975 by Data Courier, Inc. Used with permission of Data Courier, Inc.

Figures 50--51: From *Science Citation Index*, copyright 1973 by the Institute for Scientific Information. Used with permission of ISI.

Figures 52-54: From *Current Contents/Life Sciences*, vol. 19, copyright 1976 by Institute for Scientific Information. Used with permission of ISI.

Figures 55-56: From *The Use of Biological Literature*

by Robert T. Bottle. Copyright 1971, by Butterworths & Co. Used with permission of Butterworths & Co.

Figure 57: From *Guide to Reference Books*, 9th ed. by Eugene P. Sheehy. Copyright 1976 by the American Library Association. Reprinted by permission of the American Library Association.

Figure A: United States Library of Congress card.

Figure B: From *Readers' Guide*, 1975-1976. Copyright 1975, 1976. Used with permission of The H.W. Wilson Company, publisher.

Figures C-H: From *Chemical Abstracts*, 1962-66; 1967-71; 1972; 1977. Copyright 1967, 1972, 1977 by the American Chemical Society. Used with permission of the American Chemical Society.

Figures I-M: From *Zoological Record*, 1971. Copyright 1974 by Zoological Society of London. Used by permission of the Zoological Society.

Introduction

The Frustrations of a Term Paper

"A term paper will be due the last week of the course." If you are like most students, when you hear these words on the first day of class, you devoutly wish you had taken another course. It is not that you are a loafer. You just know from previous experience that, to get a decent grade on a term paper, you will have to cope once again with that monument to frustration: your college or university library. Never yet have you managed to find those two essentials: a topic that really captured your interest and the books and articles that both stimulated and satisfied your curiosity about the topic. Sometimes you came close, but not without spending hour after frustrating hour thumbing through the card catalog and browsing in the stacks.

Our Purpose and Method

Your task may never be easy, but it will be much easier if you learn to use an effective search strategy and the appropriate reference sources. That is what this book is all about.

"But what is search strategy?" you ask. For our purposes search strategy may be briefly defined as a systematic way of finding an appropriate term-paper topic and then finding enough important library materials on that topic. As this book will show, search strategy involves much, much more than just looking up a few titles in the card catalog.

This book uses examples to teach basic search strategy and reference sources, which is not a bad plan if it is true that "the three best ways to teach are by example, by example, and by example." The primary example used throughout this book is a typical term-paper topic. The topic chosen is most relevant to Ecology courses. Excerpts from the basic reference sources relate to the sample topic and demonstrates both search strategy and the use of these sources. This teaching method gives you a concrete demonstration which you can readily adapt to your own topic.

The Goal of this Book

Of course, learning takes time. As Mrs. Hoxie, the tennis coach, used to say, "The first 10,000 balls are the hardest." In learning the search strategy and reference sources for term papers in biology, only the first two are the hardest. After you write two term papers under the close guidance of this book, you will probably know what to do well enough to put the book aside, except for Appendix IV.

Then you can expect to see your former frustrations turn to joy. You will begin to look on future term-paper assignments as a welcome chance to pursue your own interests and perhaps be truly creative. Using this book can be the difference between term papers that are a labor of love and papers which are just plain labor.

John Hart, "B.C." comic strips. New York: Field Enterprises, 1971. By permission of John Hart and Field Enterprises, Inc.

"In stable populations, predation is beneficial to the prey as a group, even though it brings about death to certain individuals. The result of interferring in a prey--predator relationship is illustrated by the case history of the Kaibab deer in Arizona earlier in the century."

–Baker, Jeffrey J.W. and Garland E. Allen, 1967, *The Study of Biology*, Reading, Mass.: Addison--Wesley, p.485.

How to Begin to Choose a Topic

If you can choose your topic, choose a subject which really interests you. Such a topic will energize you, stir your imagination, and enliven your writing. Don't choose a topic just because it looks easy or because your professor seems to be interested in it.

You have chosen to work on the Kaibab Plateau deer population and its size in relation to predator population. You have for a long time been inter--ested in predator--prey relationships and you re--member reading about the Kaibab deer population in your general biology text (quotation above). However, other than a hazy notion of what preda--tor--prey relationships might be, and that the Kai--bab deer population is a good example of such a relationship, you know little about the topic.

Why Look for Authoritative Summaries

You are a fledgling scholar in the subject, and do not have an understanding of the scope of the sub--ject and its various components, and you know from previous library searches there are likely to be technical terms, for which you will not know the meaning. For these reasons it is important that you begin your search by looking for authorita--tive summaries of the topic. Hopefully these sum--maries will do three things:

1) Give you an overview of the topic.
2) Define commonly used technical terms.
3) Provide a bibliography of the "best" sources.

Where to Find Summaries

Summary discussions can be found in special ency--clopedias, in textbooks, in books put on reserve by your instructor, and in other books your reference librarian can recommend. The best encyclopedia to use at the beginning of a biology search is the *McGraw--Hill Encyclopedia of Science and Tech--nology*, 4th ed., 15 volumes (New York: McGraw--Hill, 1977). It is the most comprehensive up--to--date encyclopedia in the sciences. While there are specialized biological encyclopedias (discussed later and listed in Appendix IV), which are useful for particular topics, none is as generally useful as the *McGraw--Hill Encyclopedia.* In its fifteen volumes there are over 7,500 articles by leading scholars, fully indexed in the last volume. This thorough indexing makes consultation of the index a useful first step in using the encyclopedia. It is particu--larly useful for getting access to all the material on a subject that may be scattered amongst articles other than the primary one, and in cases where there are no separate articles on your topic.

The topic of predator--prey relationships is an example of the situation in which there is no separate article. There is no article in the encyclo--pedia under the heading "Predator," "Predation," or "Prey," but in the *index* under "Predation" (see FIGURE 1) you would find reference to volume four, pages 431--432 which is a section of a longer article, "Ecological Interactions" (FIG--URE 2). This section of the article describes in general terms the nature of predator--prey rela--tionships and gives a few examples, but the Kaibab deer are not mentioned. Note in FIGURE 3, that the author of the article is David E. Davis. If you choose to cite an encyclopedia article in the bib--liography of your paper, it should be cited as though it were a chapter in a book. In this case: Davis, David E. (1971). Ecological Interactions in *McGraw--Hill Encyclopedia of Science and Technology,* 3rd ed., volume 4, pages 430--432. McGraw--Hill, New York. This form is that rec--ommended by the Council of Biology Editors in their *CBE Style Manual,* 3rd edition, 1972 (Wash--ington, D.C.: American Institute of Biological Sciences.)

Encyclopedias, in addition to providing intro--ductory information, are an important source of bibliographies. In the case of FIGURE 3, D.E. Davis has listed three basic texts which he felt would be the best introductory sources for the topic "Ecological Interactions." Of these, R.L. Smith's *Ecology and Field Biology* would appear to be the most useful for your topic.

When searching any index you should not search just one or two headings. Do a thorough search under as many headings as you think are related to your topic. For example, the topic you have chosen deals with deer populations, so some heading like "Deer" might be checked. Also, more general headings like "Animal Ecology" and "Animal Populations" might also be checked if you feel the need for additional background information. FIGURE 4 shows you entries from the *McGraw--Hill Encyclopedia* index under "Deer" and "Animal Ecology." Note how the index provides detailed listings for subtopics and specific aspects of the entry and how scattered the material is amongst the volumes. It is pre--cisely for these reasons that the index should be consulted.

Another encyclopedia that might be consulted for a general summary of a topic is Peter Gray's *The Encyclopedia of the Biological Sciences,* 2nd edition (New York: Van Nostrand Reinhold, 1970). This one volume encyclopedia has brief summaries of topics in all areas of biology, but has little or no information on applied biology, agriculture, medicine, and the related fields of

Precured resin finishes: fabrics
13 554
Predaceous diving beetle **3** 294;
14 445
Predation **4** 431–432
insect control, biological
7 137–138*
Predatory barracuda **2** 102
Predazzite **2** 382
Predetermined elemental time
systems **14** 633–634
Prediction: climatic **3** 188
weather *see* Weather forecasting
and prediction
Predvoditilev number **4** 210
Preemphasis (sound transmission)
5 555; **10** 123; **13** 470, 491

FIGURE 1. McGraw–Hill Encyclopedia of Science and Technolo–gy, 3rd ed., 1971. Index, p. 429.

to predict, even in the laboratory, because it depends upon many variables.

Predation. This act is the killing and eating of an individual of one species by an individual of another species. A conventional example of predation is the killing of a mouse by a cat. However, the term is applicable also to the killing of an individual by a group, such as an attack of a pack of wolves upon a deer.

Predation also refers to a process in a population. It is the statistical effect on the population by the various predators. Thus predation is said to reduce a population or to alter the age composition. In these cases the mass effect of some or all of the predators on the population is being considered.

Examples of acts of predation are commonplace.

FIGURE 2. Ecological Interaction IN McGraw–Hill Encyclopedia of Science and Technology, 3rd ed., 1971. Vol. 4, pp. 429–432.

habit has appeared independently in five families. *See* CUCKOO.

While the interaction of parasitism may be harmful to the individual, it is not necessaril harmful to the species. As for predation, the eff of parasitism may prevent overpopulation stimulate the evolution of new structures o ap-tations. A value judgment about its effe the species is precarious. *See* POPULATION E ERSAL.

[DAVID E. DAVIS]

Bibliography: S. M. Henry (ed.), *Symbiosis: Associations of Microorganisms, Plants, and Marine Organisms,* 1966; C. B. Knight, *Basic Concepts of Ecology,* 1965; R. L. Smith, *Ecology and Field Biology,* 1966.

FIGURE 3. Ecological Interaction IN McGraw–Hill Encyclopedia of Science and Technology, 3rd ed., 1971. Vol. 4, pp. 429–432.

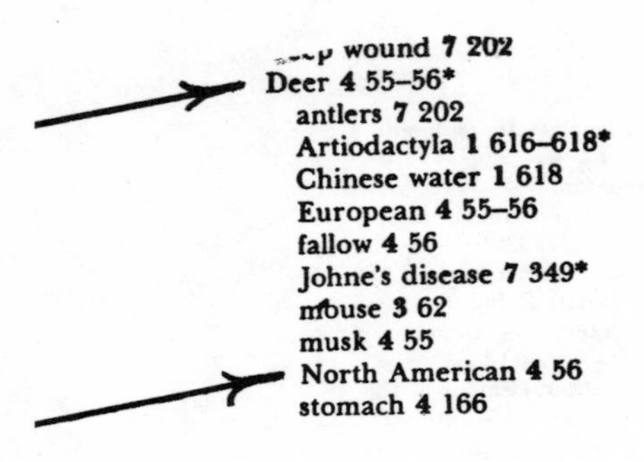

wound 7 202
Deer 4 55–56*
antlers 7 202
Artiodactyla 1 616–618*
Chinese water 1 618
European 4 55–56
fallow 4 56
Johne's disease 7 349*
mouse 3 62
musk 4 55
North American 4 56
stomach 4 166

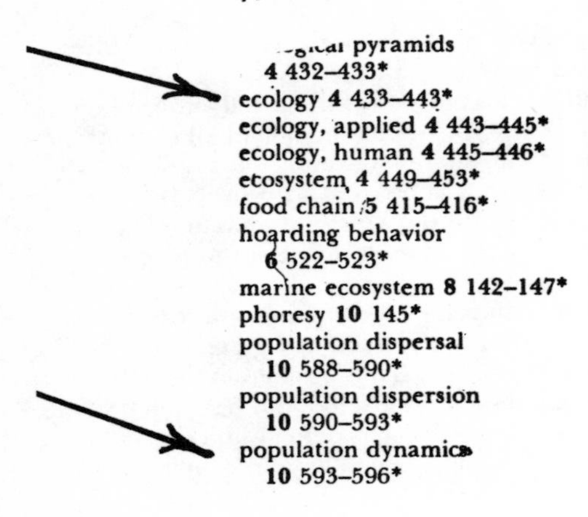

Animal dominants 3 353
Animal ecology: animal
communication 1 433–436*
animal community
1 436–439*
pyramids
4 432–433*
ecology 4 433–443*
ecology, applied 4 443–445*
ecology, human 4 445–446*
ecosystem 4 449–453*
food chain 5 415–416*
hoarding behavior
6 522–523*
marine ecosystem 8 142–147*
phoresy 10 145*
population dispersal
10 588–590*
population dispersion
10 590–593*
population dynamics
10 593–596*

FIGURE 4. McGraw–Hill Encyclopedia of Science and Technology, 3rd ed., 1971. Index, p. 27 and p. 134.

chemistry and geology. However, the articles included are well done, written by authorities, and are usually concluded with a bibliography. Unfortunately the index to Gray does not have the same depth as that of the *McGraw--Hill Encyclopedia.* Nevertheless, it is still worthwhile to check the index, as FIGURE 5 shows. The index indicates that on page 63 there is an article on Artiodactyla. From the article illustrated in FIGURE 6 you get a reference to another potentially useful source, P.L. Errington's *Of Predation and Life.*

While the *McGraw--Hill Encyclopedia* and *The Encyclopedia of the Biological Sciences* are basic sources which should be consulted at the beginning of a search for information within the general scope of biology, there are other more specialized encyclopedias useful only for study in limited areas of biology or related areas. It is not possible to remember all of them; therefore, you will want to consult one or more of three sources to identify appropriate titles:

1) Appendix IV of this book, "Basic Reference Sources for Undergraduate Biology Courses."
2) Chapter 7 of this book, "Using Guides to the Literature."
3) Your reference librarian.

For your topic, the Kaibab deer as an example of predator--prey relationships, you would probably want to consult Grzimek's *Animal Life* Encyclopedia, 13 volumes (New York: Van Nostrand

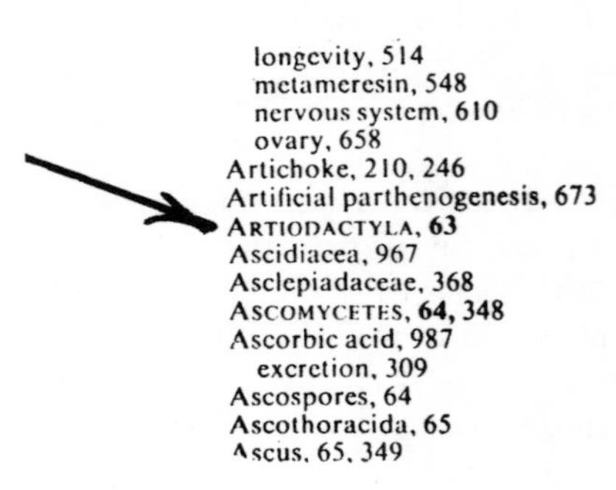

longevity, 514
metameresin, 548
nervous system, 610
ovary, 658
Artichoke, 210, 246
Artificial parthenogenesis, 673
ARTIODACTYLA, 63
Ascidiacea, 967
Asclepiadaceae, 368
ASCOMYCETES, 64, 348
Ascorbic acid, 987
excretion, 309
Ascospores, 64
Ascothoracida, 65
Ascus, 65, 349

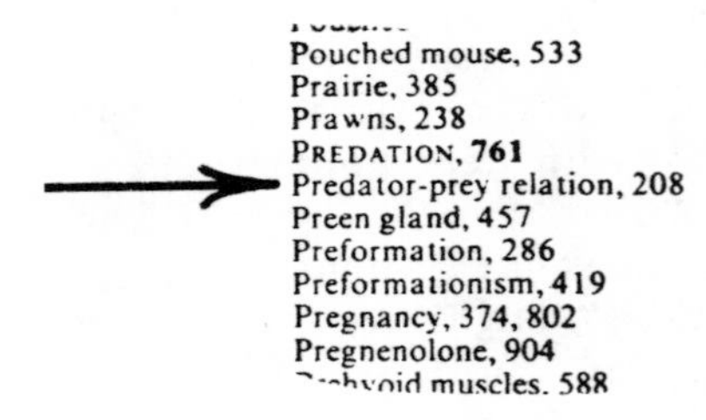

Pouched mouse, 533
Prairie, 385
Prawns, 238
PREDATION, 761
Predator-prey relation, 208
Preen gland, 457
Preformation, 286
Preformationism, 419
Pregnancy, 374, 802
Pregnenolone, 904

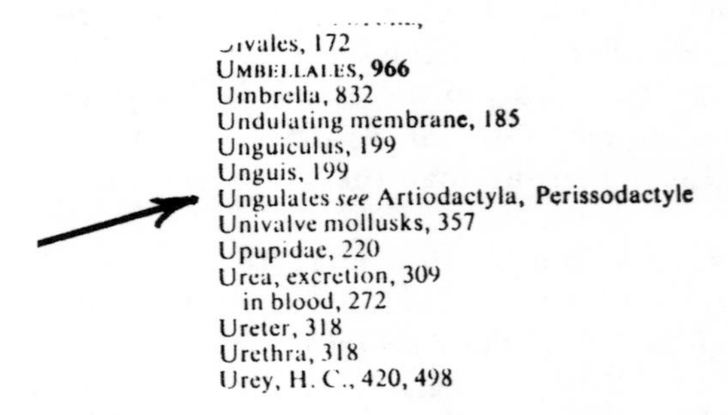

UMBELLALES, 966
Umbrella, 832
Undulating membrane, 185
Unguiculus, 199
Unguis, 199
Ungulates *see* Artiodactyla, Perissodactyle
Univalve mollusks, 357
Upupidae, 220
Urea, excretion, 309
in blood, 272
Ureter, 318
Urethra, 318
Urey, H. C., 420, 498

FIGURE 5. Gray, Peter. Encyclopedia of the Biological Sciences, 2nd ed., 1971. Index, p. 1003, 1020, and 1026. "Artiodactyla" is the scientific name of the order to which the deer belong. "Ungulates" is the general name for all hoofed animals.

PREDATION 761

PREDATION

The term "predation," derived from a word meaning "to catch," refers to the biological process resulting from the capture (and generally death) of members of a population. The individual that catches is the predator and the individual that is caught is the prey. The extent of predation describes the effect of the process on the population and for example may be "heavy" or "light." Usually the word is restricted to vertebrates and insects; for example the predation by foxes on ...nd by wasps on caterpillars. However th...

..., of the pest. Thus a balance is achie... ...iceable damage to the strawberry plants.

The principles of predation apply to diseases and parasites of humans. Indeed some of the best examples of predator-prey relations occur in measles, jungle yellow fever, and the encephalitides. However man can alter the course of the relationship either to his advantage or disadvantage according to his wisdom.

DAVID E. DAVIS

Reference

Errington, P. L., "Of Predation and Life," Ames, Iowa, State University Press, 1967.

FIGURE 6. Predation IN Peter Gray, The Encyclopedia of the Biological Sciences, 2nd ed., 1970, pp. 761–762.

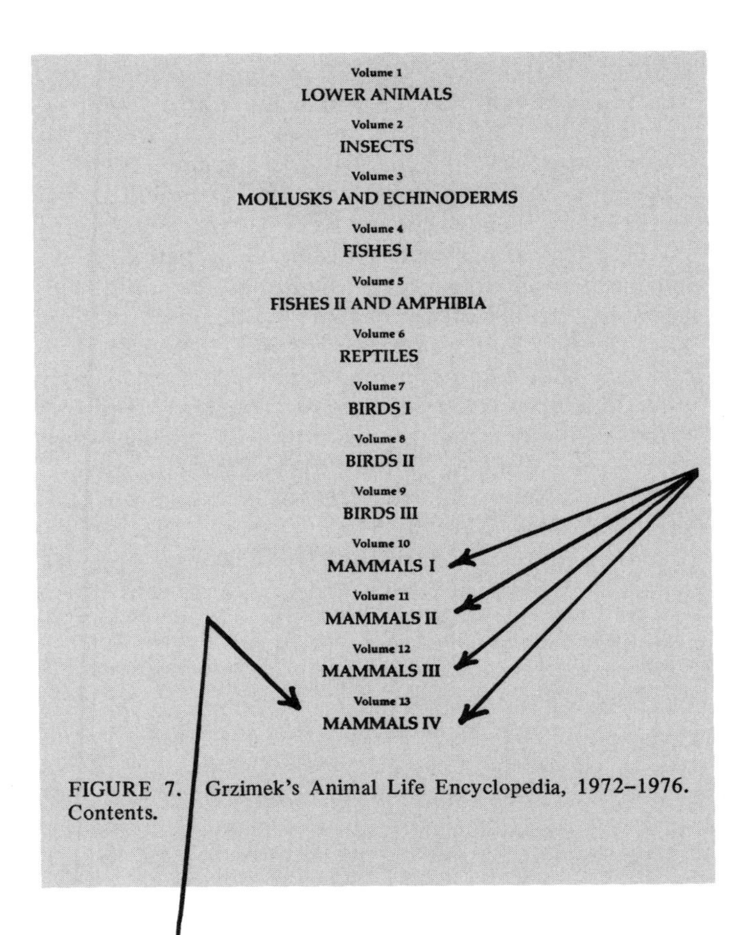

Volume 1
LOWER ANIMALS
Volume 2
INSECTS
Volume 3
MOLLUSKS AND ECHINODERMS
Volume 4
FISHES I
Volume 5
FISHES II AND AMPHIBIA
Volume 6
REPTILES
Volume 7
BIRDS I
Volume 8
BIRDS II
Volume 9
BIRDS III
Volume 10
MAMMALS I
Volume 11
MAMMALS II
Volume 12
MAMMALS III
Volume 13
MAMMALS IV

FIGURE 7. Grzimek's Animal Life Encyclopedia, 1972–1976. Contents.

Reinhold, 1972--76), and *The Larousse Encyclopedia of Animal Life,* (New York: McGraw-Hill, 1967). Grzimek does not follow the traditional arrangement of an encyclopedia but has divided the animals into groups (FIGURE 7). The volumes contain six to ten chapters each dealing with a particular group of animals. By checking the Tables of Contents of volumes 10 through 13, you could determine that volume 13 (FIGURE 8) includes one chapter on deer, and volume 12 covers the carnivores which are the deer's predators. However, since you are not informed about the Kaibab deer or its predators yet, you will have to pass up reading in these chapters for right now. You will no doubt want to return to them a little later in your search.

Several important features of Grzimek should be mentioned before you conclude this stage of the search. The encyclopedia has an excellent index for individual organisms, using both common and scientific names. For example, it is possible to pinpoint information on an organism quite easily once you know which volume covers the group to

CONTENTS

For a more complete listing
of an al names, see systematic classification or the index.

FIGURE 8. Grzimek's Animal Life Encyclopedia, 1972–1976. Table of Contents, vol. 13.

which it belongs. FIGURE 9 shows sections of the index of volume 13 which illustrates some of the ways you might find information on deer. A second noteworthy feature of Grzimek is the "Systematic Classification" which is included at the end of each volume. Here all the organisms covered in a particular volume are listed in a systematic arrangement. While it may not be apparent to you now that the systematics of the deer is something you want to know, it is generally the case that when studying a group of organisms, questions about the identity of the group and its relationship to other groups will arise during your search. By glancing at the "Systematic Classifica-tion" in volume 13 (FIGURE 10), you can quickly establish that your "deer" is a Family (Cervidae) and that there are several subfamilies only three of which are found in North America, and right now

INDEX 559

Ceratotherium simum (Square-lipped rhinoceros), 35, 37*, 49,

simum (Southern square-lipped rhinoceros), 64, 65*, 68
Cervavitus tarakliensis, 156
Cervidae (Deer), 153, **154ff**
Cervinae (True deer), 156, 160, 161f
Cervus (Red deer), 169, 175ff
– *albirostris* (Thorold's deer), **183**, 183m, 218*
– *duvauceli* (Barasingha), **172f**, 213*
– – *duvauceli* (North Indian bara-singha), 173m, 213*
– – *branderi* (Central Indian bara-singha), 173m
– – *schomburgki* (Schomburgk's deer), 172f, 173m, 213*
– *elaphus* (Red deer), 163/164*, 166, 175, 175m
(Hangul), 177

jak reevesi), 160f, 209*
Chinese roe deer (*Capreolus ca-preolus bedfordi*), 196
Chinese water deer (*Hydropotes inermis*), 194f, 195m, 219*
Choeropsis liberiensis (Pigmy hippo-potamus), 109, 110ff, 110m, 111*
Chorley, 306
Cinclus cinclus (Water black bird), 463*
oceras, 246
121f

414/415
David, Armand (Père David),
Decennatherium, 246
Deer (Cervidae), 149, 153, **154ff**
Defassa waterbuck (*Kobus ellipsi-prymnus defassa*), 402* 41 422
Dehm, 74
cour, Jean, 161, 195

FIGURE 9. Grzimek's Animal Life Encyclopedia, 1972–1976, vol. 13. Index, pp. 559–560.

SYSTEMATIC CLASSIFICATION 523

Tribe Roe Deer (Capreolini) 196
Genus (*Capreolus*) 196
Roe Deer, *C. capreolus* (Linné, 1758) 196
European Roe Deer, *C. capreolus capreolus* (Linné, 1758) 196
Siberian Roe Deer, *C. capreolus pygargus* (Pallas, 1771) 196
Chinese Roe Deer, *C. capreolus bedfordi* Thomas, 1908 196
Tribe American Deer (Odocoileini) 202
Genus *Odocoileus* 202
Subgenus *Odocoileus* (more specifically): 202
White-tailed Deer, *O. (Odocoileus) virginianus* (Zimmermann, 1780) 202
Virginian White-tailed Deer, *O. (Odocoileus) virginianus virginianus* (Zimmermann, 1780) 202
Key's White-tailed Deer, *O. (Odocoileus) virginianus clavium* Barbour and G. M. Allen, 1922) ◊ 202
Mule Deer, *O. (Odocoileus) hemionus* (Rafinesque, 1817) 204
Rocky Mountain Mule Deer, *O. (Odocoileus) hemionus hemionus* (Rafinesque, 1817) 204
Black-tailed Deer, *O. (Odocoileus) hemionus columbianus* (Richardson, 1829) 204
South American Marsh Deer, *O. (Odocoileus) dichotomus* (Illiger, 1811) 205
Subgenus *Blastoceros*: 206
Pampas Deer, *O. (Blastoceros) bezoarticus* (Linné, 1766) ◊ 206
Patagonian Kamp Deer, *O. (Blastoceros) bezoarticus color* Cabrera, 1943 206
Genus Guemals (*Hippocamelus*) 226
Peruvian Guemal, *H. antisiensis* D'Orbigny,

Veretschagin, 1955
Yellowstone Moose, *A. alces shirasi* Nelson, 1914 23
Subfamily Reindeer (Rangiferinae) 238
Genus Reindeer (*Rangifer*) 238
Reindeer, *R. tarandus* (Linné, 1758) 238
North European Reindeer, *R. tarandus tarandus* (Linné, 1758) 238
Spitsbergen Reindeer, *R. tarandus platyrhynchus* (Vrolik, 1829) 238
Eurasian Tundra Reindeer, *R. tarandus sibiricus* Murray, 1866 238
West Canadian Woodland Reindeer or Caribou, *R. tarandus caribou* (Gmelin, 1788) 238
Barren Ground Caribou, *R. tarandus arcticus* (Richardson, 1829) 238
East Canadian Woodland Reindeer, *R. tarandus sylvestrix* (Richardson, 1829) 244
Family Giraffes (Giraffidae) 247
Subfamily Forest Giraffe (Okapiinae) 24

FIGURE 10. Grzimek's Animal Life Encyclopedia, 1972–1976, vol. 13. Systematic Classification, pp. 523–524. The number to the right of each name gives the page on which the organism is discussed in the text.

you are only interested in the genus *Odocoileus*.

This otherwise excellent source has only one serious weakness; it lacks an adequate bibliography. While there is a list of references at the conclusion of each volume, it is not subdivided by organism groups, with the result that to use it requires scanning the entire list. Furthermore you cannot always count on finding sources that are useful for your particular topic or organism.

Why Ask Your Reference Librarian

Maybe encyclopedias, textbooks, and reserve books will all fail to supply the needed summaries. Does this mean you should change topics? Not necessarily. First ask your reference librarian for help. A reference librarian's job is to help students use the library effectively, but s/he is not a mind reader. So, if you have a problem, go to the Reference Desk. Ask your question as precisely as you can and tell the librarian where you have already looked. S/He may lead you to encyclopedias you never heard of, or s/he may show you particular subject headings in the card catalog. However, after some looking around, s/he may conclude that the library does not have the summary discussions you need. In fact, s/he may recommend that you change your topic because the library's resources appear to be too limited. It is better to change a dead-end topic early, before you have invested too much time in it. In any case, it is always wise to talk early and often with the term paper writer's best friend--a librarian.

Your professor is also a valuable resource person who can suggest materials that are on a given topic.

To conclude this chapter, we can summarize your search up to this point as follows:

1) You have used the *McGraw--Hill Encyclopedia of Science and Technology* and Peter Gray, *The Encyclopedia of the Biological Sciences,* to locate background material on deer and on the nature of predator--prey relationships.
2) From these encyclopedias you have identified two sources that can be consulted for more information: R.L. Smith, *Ecology and Field Biology,* and P.L. Errington, *Of Predation and Life.*
3) You consulted specialized encyclopedias, i.e. Grzimek and Larrouse, for additional information on deer. These sources were identified from the listings in Appendix IV of this book and the sources discussed in chapter 7 of this book.

Limitations and Difficulties of the Card Catalog

You may recognize the card catalog as the most important single reference source in the library, but are you aware of its limitations? It indexes only the *general* subject of *books*. It does not index parts of most books, nor does it provide access to periodical articles. In most libraries the card catalog does not include government documents. And it does not give much help in evaluating the books it lists.

A card catalog is usually simple to use if you need a particular book and know its author or title. You just look it up and copy its call number. The big difficulty with the card catalog comes when you try to find what books the library has on a particular subject. Then you must cope with the special language of subject headings, which is significantly different from spoken English.

Using What You Already Know

One way to identify the proper headings for your topic is to use subject tracings. Earlier in this text we searched the encyclopedias and were given a reference to R.L. Smith's *Ecology and Field Biology*. Naturally, you would go to the library's card catalog to see if your library has it. When you look it up note the subject tracing "Ecology" printed at the bottom of the card (FIGURE 11)

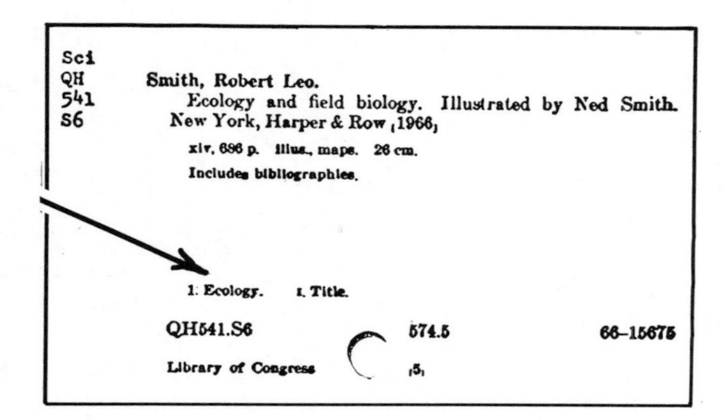
Sci
QH
541
S6

Smith, Robert Leo.
Ecology and field biology. Illustrated by Ned Smith. New York, Harper & Row [1966]
xiv, 686 p. illus., maps. 26 cm.
Includes bibliographies.

1. Ecology. I. Title.

QH541.S6 574.5 66-15675

Library of Congress [5]

FIGURE 11. Catalog card for R. L. Smith, Ecology and Field Biology.

and then look for other books under that same heading.

When you get to that point in the card catalog, you will find a number of books, and you will need to select the few "best" books on the subject. But how can you do that without looking at the books themselves? While there certainly is no way to identify the best books from the cards alone, there is information on the cards that can help eliminate books that would most likely *not* be useful for the study of your topic. FIGURE 12 shows you several cards which are filed under the heading "Ecology." These illustrate the use of the following criteria which help make tentative judgments about which books to skip:

1) *Date of publication.* The most recent books are generally more desirable because their authors have had access to more accumulated knowledge. This is particularly true in the sciences where new research is constantly requiring a reassessment of established ideas and models of how life functions. For example, ecology has developed so much in the last fifteen years that it is wise to avoid most books in that field published before 1960.
2) *Author's authority.* You may instantly recognize certain authors as being authorities, perhaps your professor put one of their books on reserve or because you remember their names from one of the selective bibliographies.
3) *Bibliographical note.* A book with a bibliography will generally be more useful and scholarly than one published without a bibliography.
4) *Publisher's reputation.* The major university presses, such as the University of Chicago Press, as well as independent publishers such as Harper and Row, Saunders, Wiley, and Academic, are among the publishers which can be trusted to publish worthwhile biology books.
5) *Edition number.* Generally, a book has proved its value if it has been published as revised, enlarged, or numbered (i.e.

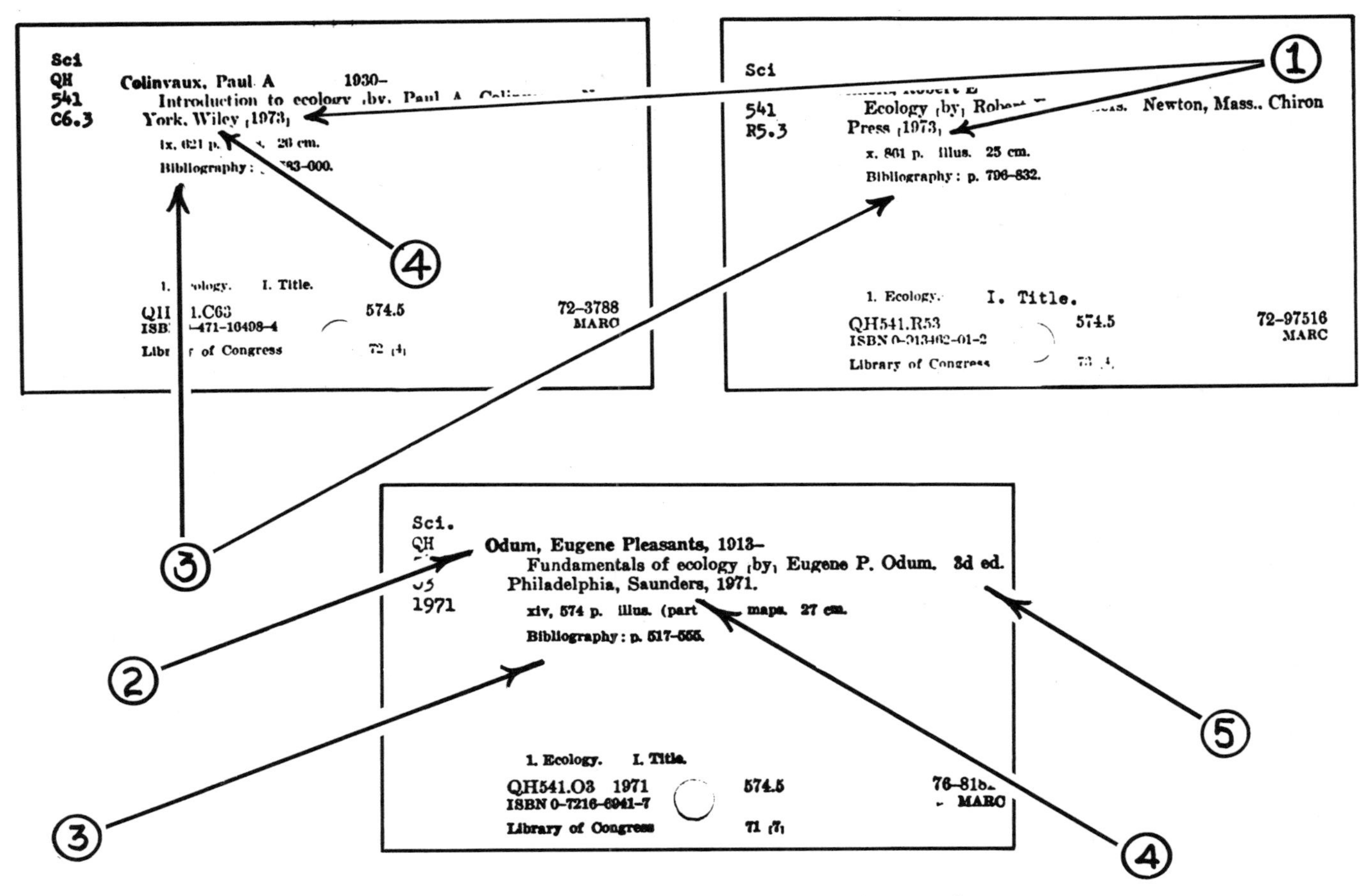

FIGURE 12. Catalog cards under the heading "Ecology."

FIGURE 13. John Hart, "B.C." comic strip. New York: Field Enterprises, 1971. By permission of John Hart and Field Enterprises, Inc.

third) editions.

The Language of Subject Headings

The language of subject headings is arbitrary. For example, the phrase, "the ecology of animals," is rendered as "Zoology--Ecology" in the language of subject headings. It is not "Animals--Ecology," "Ecology of Animals," or any of a number of other possibilities.

You cannot count on logic to give you the correct subject heading. For instance, if you learn that the language of subject headings decrees the use of "Animal communication" and "Plants--Evolution," you cannot deduce from these examples that "Animal ecology" or "Animals--Evolution" will be used and in fact they are not used. The established headings are: "Zoology--Ecology," and "Evolution." Nevertheless, making the effort to know the correct subject heading is essential. If you are to get the right response in your dialog with the catalog, you need to learn its vocabulary. Therefore, when the use of tracings described above does not produce satisfactory results, you will have to identify your own headings.

Most individual libraries do not invent their own subject headings, but use whatever headings are assigned to books by the nation's largest library, the Library of Congress. Fortunately, the Library of Congress has published a dictionary that clarifies the language of subject headings, which is: *Library of Congress Subject Headings,* 8th edition (Washington: Library of Congress, 1975). These big red volumes and their paperback supplements are a complete guide to the subject headings and cross references used in the card catalog. These books are essential because subject headings, besides being arbitrarily chosen, are extremely numerous and often changed (see FIGURE 14). Most libraries put "see" and "see also" cards in their catalogs as cross references, but they cannot always keep up with the large number of additions and changes in subject headings. For this you need the subject heading book and its supplements, but you may have trouble finding them. Some libraries do not put them out, because they were prepared primarily for catalogers. You may need to ask your reference librarian to get you the volumes and show you how to use them. If they are available, but are without any sign or person to teach you how to use them, study FIGURE 14.

NOTE: There are few names of organisms included in the *Subject Heading* list. Therefore, when your topic concerns a specific plant or animal you will not get much help from the *Subject Heading* list. Be aware that like other headings there is no consistency in the type of heading;

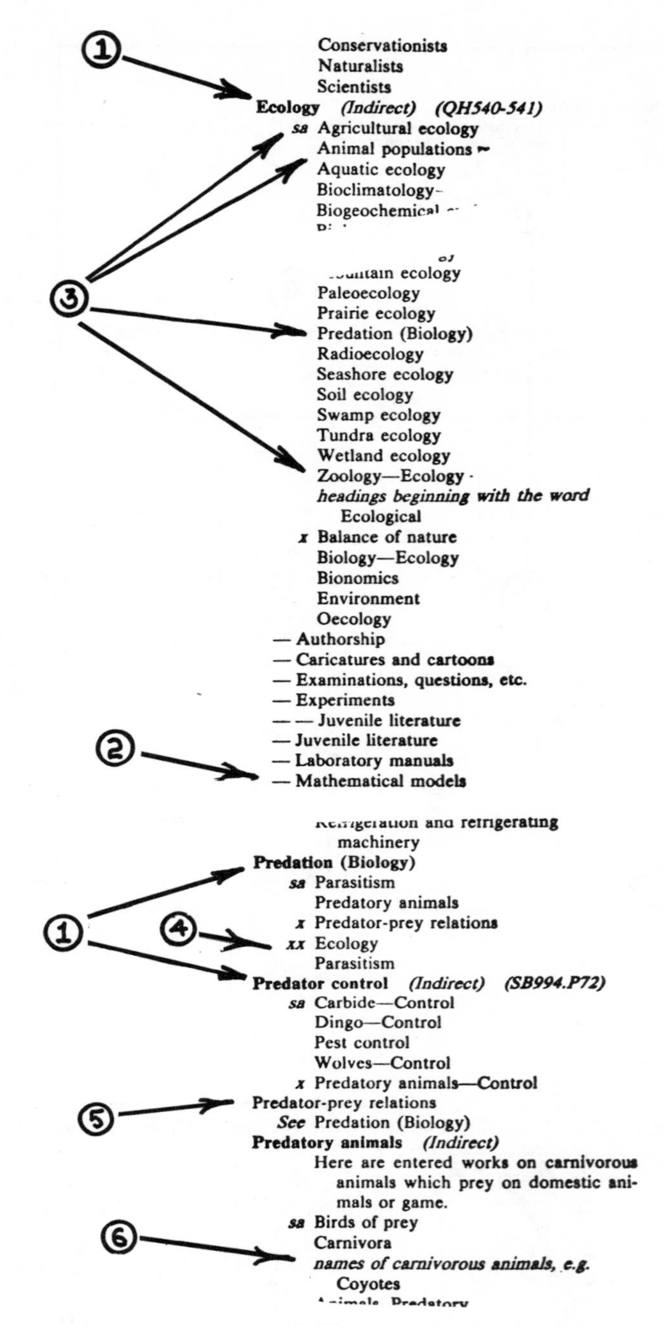

Conservationists
Naturalists
Scientists
Ecology *(Indirect)* *(QH540-541)*
sa Agricultural ecology
Animal populations
Aquatic ecology
Bioclimatology
Biogeochemical
...ountain ecology
Paleoecology
Prairie ecology
Predation (Biology)
Radioecology
Seashore ecology
Soil ecology
Swamp ecology
Tundra ecology
Wetland ecology
Zoology—Ecology
headings beginning with the word Ecological
x Balance of nature
Biology—Ecology
Bionomics
Environment
Oecology
— Authorship
— Caricatures and cartoons
— Examinations, questions, etc.
— Experiments
— — Juvenile literature
— Juvenile literature
— Laboratory manuals
— Mathematical models

Refrigeration and refrigerating machinery
Predation (Biology)
sa Parasitism
Predatory animals
x Predator-prey relations
xx Ecology
Parasitism
Predator control *(Indirect)* *(SB994.P72)*
sa Carbide—Control
Dingo—Control
Pest control
Wolves—Control
x Predatory animals—Control
Predator-prey relations
See Predation (Biology)
Predatory animals *(Indirect)*
Here are entered works on carnivorous animals which prey on domestic animals or game.
sa Birds of prey
Carnivora
names of carnivorous animals, e.g. Coyotes

FIGURE 14. Library of Congress Subject Headings, pp. 554 and 1438.

1. Headings in bold print are authorized for use.
2. Dashes indicate subject subdivisions. Under "Ecology," the most useful subject subdivision for our purpose is "Mathematical Models." Six other subdivisions of "Ecology" are shown and one of these – "Experiments" – is further subdivided.
3. "sa" (standing for "see also") refers to one or more related headings. The "see also" reference under "Ecology" which are most relevant to our needs are "Animal Populations," "Predation (Biology)," "Zoology–Ecology." Note that seventeen other "see also" references are also shown in the same column.
4. "xx" (standing for "see also") refers to related headings which are more general in scope than the heading under which they are listed. In this example, "Ecology" is listed as a more general heading under "Predation." These headings should be used only when the specific heading does not lead to anything useful in the card catalog.

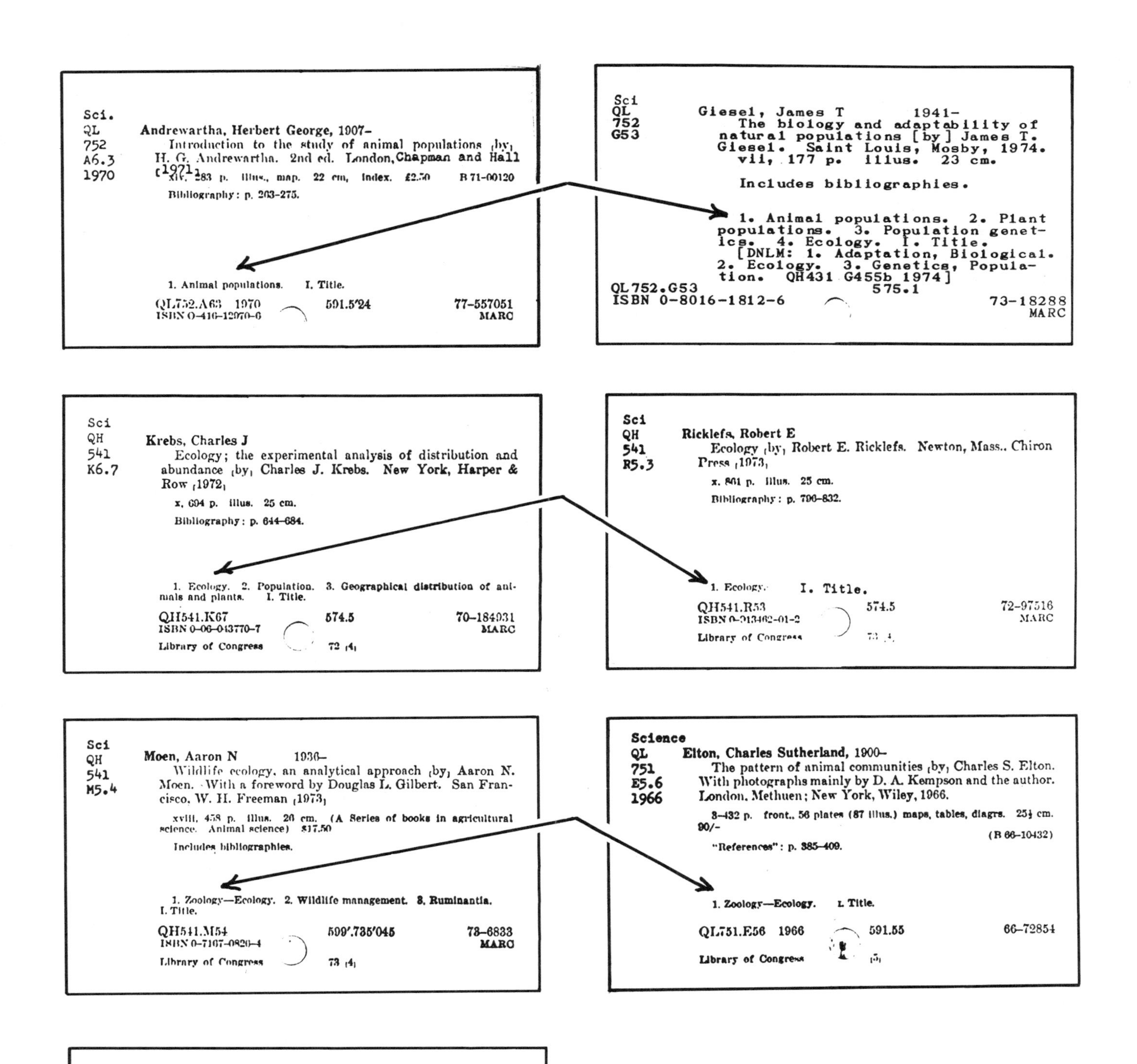

Sci.
QL
752
A6.3
1970

Andrewartha, Herbert George, 1907-
Introduction to the study of animal populations [by] H. G. Andrewartha. 2nd ed. London, Chapman and Hall [1971]
xiv, 283 p. illus., map. 22 cm. index. £2.50 B 71-00120
Bibliography: p. 263-275.

1. Animal populations. I. Title.
QL752.A63 1970 591.5'24 77-557051
ISBN 0-416-12970-6 MARC

Sci
QL
752
G53

Giesel, James T 1941-
The biology and adaptability of natural populations [by] James T. Giesel. Saint Louis, Mosby, 1974.
vii, 177 p. illus. 23 cm.
Includes bibliographies.
1. Animal populations. 2. Plant populations. 3. Population genetics. 4. Ecology. I. Title. [DNLM: 1. Adaptation, Biological. 2. Ecology. 3. Genetics, Population. QH431 G455b 1974]
QL752.G53 575.1
ISBN 0-8016-1812-6 73-18288
MARC

Sci
QH
541
K6.7

Krebs, Charles J
Ecology; the experimental analysis of distribution and abundance [by] Charles J. Krebs. New York, Harper & Row [1972]
x, 694 p. illus. 25 cm.
Bibliography: p. 644-684.

1. Ecology. 2. Population. 3. Geographical distribution of animals and plants. I. Title.
QH541.K67 574.5 70-184931
ISBN 0-06-043770-7 MARC
Library of Congress 72 [4]

Sci
QH
541
R5.3

Ricklefs, Robert E
Ecology [by] Robert E. Ricklefs. Newton, Mass., Chiron Press [1973]
x, 861 p. illus. 25 cm.
Bibliography: p. 796-832.

1. Ecology. I. Title.
QH541.R53 574.5 72-97516
ISBN 0-913462-01-2 MARC
Library of Congress 73 [4]

Sci
QH
541
M5.4

Moen, Aaron N 1930-
Wildlife ecology, an analytical approach [by] Aaron N. Moen. With a foreword by Douglas L. Gilbert. San Francisco, W. H. Freeman [1973]
xviii, 458 p. illus. 26 cm. (A Series of books in agricultural science. Animal science) $17.50
Includes bibliographies.

1. Zoology—Ecology. 2. Wildlife management. 3. Ruminantia. I. Title.
QH541.M54 599'.735'045 73-6833
ISBN 0-7167-0820-4 MARC
Library of Congress 73 [4]

Science
QL
751
E5.6
1966

Elton, Charles Sutherland, 1900-
The pattern of animal communities [by] Charles S. Elton. With photographs mainly by D. A. Kempson and the author. London, Methuen; New York, Wiley, 1966.
3-432 p. front., 56 plates (87 illus.) maps, tables, diagrs. 25½ cm. 90/-
(B 66-10432)
"References": p. 385-409.

1. Zoology—Ecology. I. Title.
QL751.E56 1966 591.55 66-72854
Library of Congress [5]

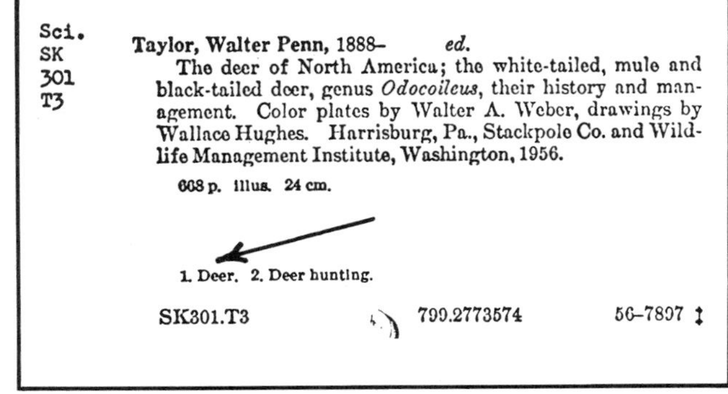

Sci.
SK
301
T3

Taylor, Walter Penn, 1888- ed.
The deer of North America; the white-tailed, mule and black-tailed deer, genus *Odocoileus*, their history and management. Color plates by Walter A. Weber, drawings by Wallace Hughes. Harrisburg, Pa., Stackpole Co. and Wildlife Management Institute, Washington, 1956.
668 p. illus. 24 cm.

1. Deer. 2. Deer hunting.
SK301.T3 799.2773574 56-7897 ‡

FIGURE 15. Catalog cards.

an organism might be listed under its common name, scientific name, and/or the field of study which gives attention to the group (e.g. Birds, Aves, Ornithology).

It appears from a study of the subject heading list that the following headings might be helpful for your topic predator--prey relationships of the Kaibab deer:

"Animal Populations"
"Predation (Biology)"
"Zoology--Ecology"
"Deer" (or some other appropriate term, i.e. *Odocoileus*, Cervidae, Ungulates)

You might wonder why you need so many headings. Is not one of them going to be better than the rest? Because there is no heading which exactly represents your topic, it is important to select several headings from the subject heading list and check each of them to see which are most useful. In other words, while a search of any one heading may be successful, no term would be complete, and therefore you might miss important material, unless all four headings are checked. FIGURE 15 shows the cards for books that appear to be the most useful.

*   *   *

Summary

1. The Card Catalog is limited because it indexes the *general* subjects of *books* only, and it does not evaluate them.
2. The language of subject headings is complex, and there are inconsistencies. The best guide to subject headings is *Subject Headings,* two large red volumes with supplements. Subject headings preceded by *sa* and *xx* are authorized for use in the card catalog, but headings preceded by "x" are not.
3. Use headings that have been applied to books about which you already know.
4. Search in *Subject Headings* for the several headings which best represent or encompass your topic. Usually you will be using headings that are more general than your topic.
5. If a heading produces insufficient material, use a more general heading.
6. You can tentatively evaluate a book on the basis of its catalog card by noting the following: date of publication, author's authority, publisher's reputation, bibliographic note, and edition number.

Once you have used the card catalog and have identified several useful tertiary* sources, it is important that you take time to study these sources carefully rather than going ahead to look for journal articles. This pause is important to your search, because you may need to redefine your paper's scope or change topics for some reason that you were not aware of when you first selected yours. In the sources you have found thus far you should look for three things:

1) Information about your topic. Particularly information that helps you clarify its scope.
2) References to journal articles or review articles that are pertinent to your topic.
3) Important terms (the vocabulary of the field) which will be important later when you are using periodical indexes.

When you first approach a source you have identified through the card catalog take just a few seconds to get familiar with it. Look at the Table of Contents, skim the Introduction, and thoroughly examine the index. As you consult the index use specific terms which represent your topic. For example, FIGURE 16 shows a portion of the index of R.L. Smith's *Ecology and Field*

----

*There are three broad categories of scientific literature: 1) Primary literature consists of reports of research written by the research worker. These usually appear in periodicals but may also appear in books; 2) Secondary literature is reviews of a number of research articles. These reviews are surveys which provide an overview of recent developments; 3) Tertiary literature is those sources which are written for a specific purpose and audience and are designed to be instructive. This category includes textbooks, and reference tools (i.e. encyclopedias, dictionaries, handbooks and bibliographies). Their intent is not to review a specific body of literature (research articles, and reviews), but rather to summarize the most important facts and theories about a general subject.

...gulation of, 30...
of soil fauna, 264, 26...,
study of animal, 636–661
Population ecology, 5, 325–341
Population regulation, 363–382
cannibalism, 416–417
cycles, 379–382
density-dependent, 367–379
density-independent, 363–366
deterioration in quality, 378–379, 381
endocrine balance, 377–378
exploitation, 417–418
genetic feedback ...
home ...
..., 377–378
...ominance and hierarchy, 373–374
territory, 372–373
Populations, relationships between, 397–425
amensalism, 348–399
disease, *402–404*
intraspecific competition, *418–424*
mutualism, 397–398
parasitism, 399–*402*

predation, 404–418
predator-prey systems, 406–407
See also Parasitism, Parasites
Populations, sampling of, 636–637

Predation, *404–418*
community viewpoint, 413
compensatory, 411, 415
density dependence in, 407–410
functional response in, 407–409
by fungi, 268
numerical response in, 409
regulatory effect of, 413–415
in soil community, 267
theory of, 405
total response, 410–411
types of, 410–411
variables influencing, 411–413
and vertebrate populations, 411–412
Predator, facultative, 413

FIGURE 16. Smith, R.L. 1966. Ecology and Field Biology. Index, p. 600.

The life equation is a modification of the life table. The life table is a

TABLE 20-4. LIFE EQUATION OF A BLACK-TAILED DEER POPULATION ON A 36,000-ACRE AREA FROM 1949 TO 1954

TYPE OF GAIN OR LOSS	MALES			FEMALES			
	Adults	Yearlings	Fawns	Fawns	Yearlings	Adults	Total
1954 Prefawning population	151	130	153	184	161	771	1,568

Adapted from Brown, 1961

Life equations

A picture of the limitations of the growth of a population, seasonal gains and losses, and other important events occurring throughout the year can be summarized in a life-equation table (Table 20-4). Since slight changes in reproduction, survival, or sex ratios can influence the rate of increase considerably from

Predation

No phase of population interaction is more misunderstood (or hotly debated, especially by sportsmen) than predation. Predation in natural communities is a step in the transfer of energy. It is commonly associated with the idea of the strong attacking the weak, the lion pouncing upon the deer, the hawk upon the sparrow. But this idea must be modified, for predation grades into parasitism versa. Between the two exists the broad gray area of the parasitoid times is called parasitism and sometim

Theory of predation

The influence of predation on population growth of a species received the attention of two mathematicians, Lotka (1925) and Volterra (1928). Separately they proposed formulas to express the relationship between predator and prey populations. They attempted to show that as the predator population increased, the prey decreased to a point where the trend was reversed and oscillations w

Functional response

In general the predator or the parasite will take or affect more of the prey, as the density of the prey increases. However, there is also a tendency for the number of prey taken or the number of hosts affected to increase in less than a linear proportion to the total number of host or prey available (Fig. 21-3). The

Numerical response

As the density of the prey increases, the numbers of predators may also increase. But the response is not immediate. There is necessarily a time lag between the birth and the appearance of an active predator, and this time lag may prevent the predator or the parasite population from catching up with the prey. As a result

Predation and vertebrate populations

Some modification in the relationship between predator and prey exists when predation is compensatory. In this situation several variables influence predation. They include natural food preferences of the predator, physical condition of the prey and escape facilities, the abundance of alternate foods, and the density of both the prey and the predator populations (Leopold, 1933). These variables can be examined from two viewpoints, that of the prey and the predator.

Prey. The many variables that make up the odds that an animal will be captured by a predator, the prey risk, is determined in part by the availability of

FIGURE 17. Smith, R. L. 1966. Excerpts from Ecology and Field Biology, pp. 361 and 405 - 411.

Biology. From it you can easily identify sections that look important. The idea is to pinpoint quickly the appropriate sections. You are not interested in reading the whole book, although you do want to find general background material on predator--prey systems and information on the specific topic of Kaibab deer and their predators. Smith provides an excellent summary of the current (1966) theory of predation, and has included references to the key works of predation theory (FIGURE 17). Note in FIGURE 17 that Smith has carefully defined predation, discussed the theory of predation and the various aspects of predation. You should carefully read such sources, take notes on the content, and maintain a list of terms and references. In FIGURE 17 the major references and terms have been underlined. Note that in Smith's references only the author's name and year are given. This is the method used by biologists to make reference to a source from which information was taken. At the end of the chapter or book the references will be listed in alphabetical order by author (FIGURE 18). This

discussion of aggressive ...
ralist, **94**:343–355.
Brown, E. R.
1961 The black-tailed deer of western Washington, *Washington State Game Dept. Biol. Bull. No.* 13 (Olympia).
Brown, F. A., Jr.
Living clocks. *Science*. **130**:1535
1959 Hormonal responses to external ... birds, *Ibis*, **101**:478–496.
Leopold, A.
1933 "Game Management," Scribner, New York.
... 1–29.
Lotka, A. J.
1925 "Elements of Physical Biology," Williams and Wilkins, Baltimore.
Lowdermilk, W. C.
Conquest of the land through ... pasture.
Monographs, **21**:39–60.
Volterra, V.
1926 Variazione e fluttazioni de numero d'individiu in specie animali conviventi, *Mem. Accad. Lincei*, **2**:31–113 (translated in R. N. Chapman, 1931, "Animal Ecology," McGraw-Hill, New York.)
Waddington, C. H.
"The Strategy of the Genes," ...

FIGURE 18. Smith, R. L. 1966. Ecology and Field Biology. Bibliography, pp. 548, 557 - 558, 569.

method is described in the *CBE Style Manual*, pp. 152--153.

Once you have found one source that surveys a topic and provides you with a general description of a topic, then turn your attention to other sources and look for information specifically on your topic. FIGURE 19 shows you the index of Eugene Odum's *Fundamentals of Ecology*, and FIGURE 20 shows you the relevant section of the text and FIGURE 21 shows its bibliography. Here at last is that one summary paragraph on your topic! While it took a period of careful searching you now have an authoritative summary, complete with key reference. And in doing the search you have learned a great deal about preda--tor--prey relationships.

As you read the Odum paragraph you no doubt

marsh grass, 30 (Fig. ...
stages of, 29
"Decreasers," **390**, 418
Deep-scattering layer, 351
Deer, population eruptions of, 220
fallout of cesium-137 in, 464 (Table 17–3)
Democratic quotient, **410**
...ulation, 163–166
...
9–1), 440 (Fig. 16–4A), 441
Prairie. See Grasslands.
Predation, 113, **211**, **220–226**
by man, 223, 225–226
in insect control, 447
relation to diversity, 150–151
Predator-prey relationships, 34, 191–192, 194, 224. See also Host-parasites.
modeling of, 287–288

FIGURE 19. Odum, E.P. 1971. Fundamentals of Ecology, 3rd ed. Index, pp. 561, 570.

... population erupts or escapes density-dependent control. Deer populations are often cited as examples of populations that tend to erupt when predator pressure is reduced. A widely cited example of this is the Kaibab deer herd, which, as originally described by Leopold (1943) based on estimates by Rasmussen (1941), allegedly increased from 4000 (on 700,000 acres on the north side of the Grand Canyon in Arizona) in 1907 to 100,000 in 1924, coincident with an organized government predator removal campaign. Caughley (1970) has reexamined this case and concludes that while there is no doubt that deer did increase, overgraze, and then decline, there is doubt about the extent of the overpopulation and no real evidence that it was due solely to removal of predators; cattle and fire may have also played a part. He believes ... eruptions of ungulate population...

FIGURE 20. Odum, E. P. 1971. Fundamentals of Ecology, 3rd ed., pp. 220 - 221.

Carson, Rachel. 1962. *Silent Spring*. ... Co., Boston.
Caughley, Graeme. 1970. Eruption of ungulate populations with emphasis on Himalayan thor in New Zealand. Ecology, 51:53–72.
Lent, C. M. (ed.). 1969. Adap... ganisms. Amer. Zool., 9:269–426.
Leopold, Aldo. 1933. *Game Management.* Charles Scribner's Sons, New York.
———. 1933*a*. The conservation ethic. J. Forestry, ...
Life history ... Oceanogr., 2:62–69.
Rasmussen, D. I. 1941. Biotic communities of Kaibab Plateau, Arizona. Ecol. Monogr., 11:229–275.
Raunkaier, C. 1934. *The Life Form of Plants and Stati...*

FIGURE 21. Odum, E. P. 1971. Fundamentals of Ecology, 3rd ed. Bibliography, pp. 521, 535, 545.

12-1 PREDATOR-PREY RELATIONSHIPS

Predator-prey interactions have been studied in controlled situations (e.g., fish tanks as miniature aquatic systems) and historical ones, such as the attempts to correlate the populations of predators with populations of prey species. These populations have often been estimated by the take of trappers an...
lags between the peak abundance of the pr...
The lynx-h... ...ation by fish and turtles.
...often not present in the shallow brood marshes, but ducks living on larger lakes and rivers are susceptible.

Larger animals such as deer and elk have fewer predators than do the upland game birds. Fawns and calves can be taken by coyotes, and wolves prey on deer

and moose in areas such as northern Minnesota. Wolves prey almost exclusively on moose on Isle Royal. Very high mortality of white-tailed deer fawns owing to coyote predation has been observed in Texas, with losses of about 72% of the fawns within two m... (Cook et al. 1971). Mountain lions, grizzlies, and black be... on elk calves.

FIGURE 22. Moen, A. N. 1973. Wildlife Ecology, An Analytical Approach, pp. 226 - 228.

are disappointed to learn that Caughley's article suggests that the Kaibab deer are not a clear cut example of prey--predator interactions because other factors are likely involved. However, be--cause you have read a good deal on predator--prey relations as you worked through general sources you are in a good position to refine or redirect your search in an appropriate way. Perhaps you decide that since there is no definitive statement on the role of predators in the control of deer population size, you will explore that topic.

One other area of your topic also needs refine--ment. As long as you were concerned with the Kaibab deer you were not particularly concerned about the various species of deer. Now that you have changed your focus to the predators' role in the control of deer population, you must be concerned with the various species of deer. Here again, the early background searching is extremely helpful.

Grzimek's *Animal Life Encyclopedia*, you may remember, (FIGURE 9) contains an excellent synopsis of deer systematics. After looking that over again you decide to confine your search to North American Cervidae, which include the reindeer (*Rangifer tarandus*), the white--tailed deer (*Odocoileus virginianus*), the mule deer (*Odo--coileus hemionus*), and the moose (*Alces alces*).

FIGURES 22 and 23, a section from Arthur N. Moen's *Wildlife Ecology* (one of the books shown in FIGURE 15 under the heading Zoology--Ecology) is typical of the type of information which you will find on your topic. Much of it is scattered amidst a general discussion of predator--prey relations, but usually with key references. By collecting these pieces of scattered informa--tion and their solitary references (as in FIGURE 22), it is usually possible to amass a considerable

...ests. *J. Wildlife Management* 32(3): 615–618.
...ness, R. A., M. M. Nelson, and W. H. Longley. 1968. The effect of predator rem... on pheasant reproductive success. *J. Wildlife Management* 32(4): 683–697.
Cook, R. S., M. White, D. O. Trainer, and W. C. Glazener. 1971. Mortality of young white-tailed deer fawns in south Texas. *J. Wildlife Management* 35(1): 47–56.
Davis, J. W., and R. C. Anderson, eds. 1971. *Parasitic diseases of wild mammals.* A... ...State University Press. 374 pp.

FIGURE 23. Moen, A. N. 1973. Wildlife Ecology, An Analytical Approach. Chapter 12, Bibliography, p. 239.

The role of predators in the control of North American deer population size.

I. Introduction
   A. Rasmussen describes removal of predators as the reason for the disruption of Kaibab deer population.
   B. Caughley rejects Rasmussen's claim by pointing out the removal of cattle (herbivore competitors with the deer) and decreased numbers of fires.

II. Role of predators in control of populations of North American deer?
   A. Mule deer (what are its predators?)
   B. Reindeer (what are its predators?)
   C. Moose
      1. Predators are wolves.
      2. Isle Royal--more information?
   D. White-tailed deer
      1. Predators are coyotes (Cook, etc., etc., 1971)
   E. Elk (=)
      1. Predators are mountain lions, grizzly and black bears.

III. Key Terms
   Predation
   Predator
   Prey
   Deer (=Cervidae)
   Reindeer (=Rangifer tarandus)
   White-tailed deer (=Odocoileus virginances)
   Mule deer (= Odocoileus hemionus)
   Moose (=Alces americana)
   Elk (=)
   Coyote (=)
   Grizzly bear (=)
   Black Bear (=)
   Mountain lion (=)
   Wolf (=)
   Kaibab deer (=)
   Isle Royal

IV. Key references

Brown, E.R. 1961. The black-tailed deer of western Washington, Washington State Game Dept. Biol. Bull., No. 13 (Olympia).

Caughley, Graeme. 1970. Eruption of ungulate populations with emphasis on Himalayan Thor in New Zealand. Ecology, 51: 53-72.

Cook, R.S. et. al. 1971. Mortality of young white-tailed deer fawns in south Texas. J. Wildlife Management 35(1):47-56.

Leopold, Aldo. 1943. Deer irruptions. Wisconsin Conservation Bull., Aug. 1943. Reprint in Wisconsin Cons. Dept. Publ., 321: 3-11.

Rasmussen, D.I. 1941. Biotic communities of Kaibab Plateau, Arizona. Ecol. Monographs, 11:229-275.

FIGURE 24. Summary outline, key terms and references.

body of information on your topic.

Once you have examined all the sources identified through the card catalog, it is a good idea to begin to outline your topic and note the most important terms and references. This will help you decide 1) what direction you want to go in, 2) what additional information you need, and 3) the outline provides the terms for your searches in periodical indexes. FIGURE 24 is an example of the kind of summary outline you might produce.

You can now quickly review the indexes of sources you have already searched, and look at the additional sources you have yet to consult for information on the role of predators in the control of North American Cervidae population sizes.

*   *   *

Summary

1. After your initial search of the tertiary and secondary literature it is vital that you reevaluate the scope of your topic and make adjustments when necessary.
2. You want a topic that is manageable within the limits set by the instructor. Therefore your topic should *not* be:
   a) so broad it results in a superficial paper;
   b) so narrow that there are insufficient sources; or
   c) inappropriate because of current knowledge in the field.
3. Point "c" above was the case in this example, and you refocused your topic in response to that situation.

For two reasons tertiary sources do not often reflect the current status of a specific topic, nor will they indicate the directions in which current research is headed. First, the writing and publishing of a book is a complicated process that can take as many as three years or longer to complete. Second, the type of sources you have looked at so far were written to present the factual and conceptual scope of a broad topic (e.g. ecology, animal ecology).

To obtain a perspective on the current status of a subject and to get the latest information two types of sources should be used: review serials and research reports. We will deal with review serials here, and will cover access to research reports in Chapters 5 and 6.

Review serials may be defined as publications that come out in parts, and have a common title. These review serials contain a number of review articles; articles that give a critical analysis of recent developments in a particular area (usually the one or two years previous to publication). These publications are noted for the variety of ways in which they are issued and the number of different ways libraries will handle them. Some are monthly or quarterly journals which may be part of a library's separate journal collection or may be shelved with the library's books on the same subject. Others are annual or biennial and usually are shelved with other books on the general subject of the title (e.g. ecology, genetics, medicine, plant physiology). Still others are irregular in their publication schedule.

Since it is impossible for you to know about all of the hundreds of review serials that exist, you need someone or something to help you identify the appropriate titles. There are three possibilities listed below.

1. Consult Appendix II of this guide, and look under the appropriate subject. In this case, ecology. However, since the list in Appendix II will no doubt go out of date fairly quickly you will want to try the other methods listed below.
2. Consult the subject card catalog of your library under the broad subject heading into which your topic fits and look for the subdivisions "Periodical" and "Annuals." This can be a tricky matter and usually requires searching under more headings than you would initially think to check. For example, while ecology is the obvious heading for your topic, other possible subjects are "Zoology," and even "Biology" since there may be review serials in those areas which contain relevant articles.
3. Ask your librarian or professor to suggest possible sources.

Once you have identified the appropriate title(s) you will need to consult your library's catalog to find out if they are present. It may be that you have to check the card catalog and/or a periodicals list or a serials catalog. In any case you will search under the title of the serial, and by all means ask the librarian for help if you are having difficulty determining whether the library has the title you want. Consult the section of Appendix II which lists the review serials in ecology. Once you have determined that your library has them and where they are shelved, you can consult them for articles. Remember, you are only interested in the material that is more current than the tertiary sources you have already consulted. Because it often takes as much as two or three years to get a review published, you should begin your search with material published two or three years before publication of the recent relevant texts you used. In this case, the most useful sources were Smith (1966), Odum (1971), and Moen (1973); therefore, you should confine your search to the period 1971 to the present.

Most review serials have some form of indexing, and while there are some variations there are two basic types of subject approaches (many review serials also have an index of authors cited). One type is a cumulative list of titles (FIGURE 25) or tables of contents for all or a few of the previous volumes. Therefore, it is only necessary to check the latest volume to get some idea of the major topics covered in recent volumes. From FIGURE 25 it is easy to identify two articles that may be

Advances in Ecological Research, Volumes 1—8:

Cumulative List of Titles

Analysis of processes involved in the natural control of insects, **2**, 1
The distribution and abundance of lake-dwelling Triclads—towards a hypothesis, **3**, 1
The dynamics of aquatic ecosystems, **6**, 1
The dynamics of a field population of the pine looper, *Bupalus piniarius* L. (Lep., Geom.), **3**, 207
Ecological aspects of fishery research, **7**, 115
Ecological conditions affecting the production of wild herbivorous mammals on grasslands, **6**, 137
Ecological implications of dividing plants into groups with distinct photosynthetic production capacities, **7**, 87
Ecological studies at Lough Ine, **4**, 198
Ecology of fire in grasslands, **5**, 209
Ecology, systematics and evolution of Australian frogs, **5**, 37
Energetics, terrestrial field studies, and animal productivity, **3**, 73
Energy in animal ecology, **1**, 69
Forty years of genecology, **2**, 159
The general biology and thermal balance of penguins, **4**, 131
Heavy metal tolerance in plants, **7**, 2
Human ecology as an interdisciplinary concept: a critical inquiry, **8**, 2
Integration, identity and stability in the plant association, **6**, 84
Litter production in forests of the world, **2**, 101
The method of successive approximation in descriptive ecology, **1**, 35
Pattern and process in competition, **4**, 1
Population cycles in small mammals, **8**, 268
The production of marine plankton, **3**, 117
Quantitative ecology and the woodland ecosystem concept, **1**, 103
Realistic models in population ecology, **8**, 200
A simulation model of animal movement patterns, **6**, 185
Studies on the cereal ecosystem, **8**, 108
Studies on the insect fauna on Scotch Broom *Sarothamnus scoparius* (L.) Wimmer, **5**, 88
Soil arthropod sampling, **1**, 1
A synopsis of the pesticide problem, **4**, 75
Towards understanding ecosystems, **5**, 1
The use of statistics in phytosociology, **2**, 59
Vegetational distribution, tree growth and crop success in relation to recent climatic change, **7**, 177

FIGURE 25. Advances in Ecology Research, Cumulative list of titles, vol. 8.

relevant. But it is important not to rely exclusively on these title lists. Since you are interested in specific information on the role of deer predators in controlling deer population size, you should use the second type of subject approach, the individual volume indexes.

For example, FIGURE 26 illustrates the usefulness of the index in pinpointing specific information. FIGURE 27 shows the specific information that the index leads to and FIGURE 28 the references which the author of the review cites. These citations are one of the most important reasons why you should search for a review serial. These two references should be added to your summary outline (FIGURE 24) which we suggested you produce after studying tertiary sources. These will be most helpful in the completion of your outline. If you have done a thorough and thoughtful job you will be as knowledgeable about the subject and its bibliography as any but the most expert research worker in the field. You know the dimension of the topic, the fundamental problems still unresolved; you know who the major contributors are and what they have written. Now you are ready to plunge into the ocean of recent research on the role of predators in controlling deer population size.

...ceros (Bellastomatidae), 36
Life-forms, 266
Likelihood function, 140 ff.
Limiting conditions, 211, 242
Linanthus, 327 ff., 347
Linaria, 288, 303, 311 ff., 326
Lindera, 290, 313
Linear variable displaceme...
Liriodendron, 176 ff...
tulipifera, 17...
Loblolly ..., 175 ff.
L... ...19, 232

plant, 1...
process, 133
Moisture status of soil, 258
Molybdenum, 336
Montmorillonite, 261
Moose-wolf system, 6, 119
Mosquito, 38, 83, 95 ff., 107, 113
larvae, 162
Moth larvae, 37
Mouse, 109
Mussels (*Mytilus*), 10, 37, 65, 80, 95 ff., 102

FIGURE 26. Advances in Ecological Research, vol. 9. Subject index, p. 380.

The follow... two examples, a moose–wolf system and a barnacle–predator system, ma... ...strate the importance of the existence of an invulnerable class in the prey. The moose–wolf story may also show that predators can stabilize a herbivore population that otherwise would be unstable, though the evidence is by no means overwhelmingly convincing. The moose–wolf interaction on Isle Royale (Mech, 1966; Jordan *et al.*, 1967) is an example of that archetypal predator–prey system, the large predator and its ungulate prey. Mech records estimated and guessed fluctuations in abundance of the moose from their colonization of the island in the early 1900's until the present. There is eviden... ...population declines, emaciated carcasses, and sever...

FIGURE 27. Predation and population stability IN Advances in Ecological Research, vol. 9, pp. 6 - 7.

...associated with biological control of Klamath weed species. *J. Range Mgm.* **12**, 69–82.

Ivlev, V. S. (1961). "Experimental Ecology of the Feeding of Fishes." Yale University Press, New Haven, Connecticut, U. S. A.

Jordan, P. A., Shelton, D. C. and Allen, D. L. (1967). Numbers, turnover, and social structure of the Isle Royale wolf population. *Am. Zool.* **7**, 233–252.

Kitching, J. A. and Ebling, F. J. (1967). Ecological studies at Lough Ine. *Advances in ecol. Res.* **4**, 197–291.

Krebs, J. R. (1973). Behavioural aspects of predation. *In*: "Perspectives in Ethology" (Eds P. P. G. Bateson and Peter H. Klopfer). Plenum Press, New York.

May, R. M. (1973a). Time delay versus stability in population models with two and three trophic levels. *Ecology* **54**, 315–325.

May, R. M. (1973b). "Stability and Complexity in Model Ecosystems." Princeton University Press, Princeton, New Jersey, U.S.A.

Mech, L. D. (1966). "The Wolves of Isle Royale." *U.S. Nat. Park Serv. Fauna Nat. Parks U.S., Fauna Series* No. 7.

Messenger, P. S. (1968). Bioclimatic studies of the aphid parasite *Praon exsoletum*. I. Effects of temperature on the functional response of females to varying host densities. *Can. Ent.* **100**, 728–41.

Mook, J. H., Mook, L. J. and Heikens, H. S. (1960). Further evidence for the role of "searching images" in the hunting behavior of titmice. *Archs néerl.* ... 13, 448–465.

FIGURE 28. Predation and population stability IN Advances in Ecological Research, vol. 9. Bibliography, pp. 122 - 123.

Summary

1. Review serials are important sources of current information on a topic and should be used as a link between tertiary sources and the original research reports literature.
2. Identification of the appropriate review serials can be difficult for the typical library user. You should consult Appendix II of this guide, check the card catalog, or ask your librarian or professor.
3. Study the relevant material carefully and fit the information, key terms, and significant citations into your outline.
4. Make an outline as you proceed, fitting in all the relevant material you have studied, including key terms and significant citations.

5 Science Citation Index and the Author Approach

The long sometimes circuitous path through ency-clopedias, other tertiary sources, and reviews has paid off. By being methodical and persistent you can usually produce the kind of results that are shown in FIGURE 29. Now you are ready to start searching for reports of original research.

For this final step in your search there are two basic approaches: 1) the author approach, and 2) the subject approach. We will explain the first approach here and the second in Chapter 6.

You may have wondered why, in the earlier stages of your search, we were so interested in

The role of predators in the control of North American deer population size.

I. Introduction
- A. Rasmussen describes removal of predators as the reason for the disruption of Kaibab deer population.
- B. Caughley rejects Rasmussen's claim by pointing out the removal of cattle (herbivore competitors with the deer) and decreased numbers of fires.

II. Role of predators in control of populations of North American deer?
- A. Mule deer (what are its predators?)
- B. Reindeer (what are its predators?)
- C. Moose
  1. Predators are wolves.
  2. Isle Royal--more information?
- D. White-tailed deer
  1. Predators are coyotes (Cook, etc.; etc., 1971)
- E. Elk
  1. Predators are mountain lions, grizzly and black bears.

III. Key Terms

Predation
Predator
Prey
Deer (=Cervidae)
Reindeer (=*Rangifer tarandus*)
White-tailed deer (=*Odocoileus virginances*)
Mule deer (= *Odocoileus hemionus*)
Moose (=*Alces americana*)
Elk (=*Alces alces*)
Coyote (=*Canis latrans*)
Grizzly bear (=*Ursus arctus horribilis*)
Black Bear (=*Ursus americanus*)
Mountain lion (=*Puma concolor*)
Wolf (=*Canis lupus*)
Kaibab deer (=*Odocoileus hemionus*)
Isle Royal

IV. Key references

Brown, E.R. 1961. The black-tailed deer of western Washington. Washington State Game Dept. Biol. Bull., No. 13 (Olympia).

Caughley, Graeme. 1970. Eruption of ungulate populations with emphasis on Himalayan Thor in New Zealand. Ecology, 51: 53-72.

Cook, R.S. et. al. 1971. Mortality of young white-tailed deer fawns in south Texas. J. Wildlife Management 35(1):47-56.

Leopold, Aldo. 1943. Deer irruptions. Wisconsin Conservation Bull., Aug. 1943. Reprint in Wisconsin Cons. Dept. Publ., 321: 3-11.

Rasmussen, D.I. 1941. Biotic communities of Kaibab Plateau, Arizona. Ecol. Monographs, 11:229-275.

FIGURE 29. Summary outline, key terms, and references.

collecting references to relevant material. "Certainly" you might have said to yourself, "those are important sources for my topic. Why shouldn't I go directly to them instead of just collecting them?" There are two reasons. First, each article citation found during the early stages of your search, was only a rough jewel of information. Each needed to be put in its proper setting, particularly in relation to other articles. That is what the tertiary and secondary sources you have

Ecology (1976) 57: pp. 390–394

PREY SELECTIVITY AND SWITCHING RESPONSE OF *ZETZELLIA MALI*[1]

MIGUEL A. SANTOS

Biology Department, Queens College (CUNY), Flushing, New York 11367 USA

Abstract. The predatory behavior of the mite *Zetzellia mali* was observed as the proportion of its three prey species changed. The prey were *Aculus schlechtendali, Panonychus ulmi,* and *Tetranychus urticae.* The predator preference remained constant as the proportion of the available prey changed. Unsuccessful attempts were made to alter predator preference by training experiments.

Key words: Mites; predator-prey stability; prey selectivity; switching experiments; Zetzellia mali.

INTRODUCTION

Elton (1927) and others have pointed out that, in theory, the presence of a variety of alternative prey species should lend stability to a community. The contention is that predators exert a stabilizing effect through compensatory mortality on the prey by "switching" the greater proportion of their attack to the prey which has become the most abundant. This switching behavior may prevent any prey species from either oversaturating its environment or becoming locally extinct.

Pearson (1966), working with various carnivores by training experiments. Also, during switching experiments, availability of prey was determined.

THE EXPERIMENTAL ANIMALS

The most abundant prey of *Zetzellia mali* E. (Acarina:Stigmaeidae) on apple leaves is the apple rust mite, *Aculus schlechtendali* N. (Prostigmata:Eriophyoidae). This phytophagous mite is minute (0.2 mm long), wormlike, and found primarily on the undersurface of leaves. It has 3–5 generations during the summer (Oatman 1972, Herbert 1974).

The two other prey species used in the experi-

394 MIGUEL A. SANTOS Ecology, Vol. 57, No. 2

glasshouse red spider mite, *Tetranychus urticae* Koch (Acarina: Tetranychidae). Entomol. Exp. Appl. **6**: 207–214.

Knisley, C. B. 1969. The biology and description of *Amblyseius umbraticus* (Acarina: Phytoseiidae), an inhabitant of apple foliage in New Jersey. Ph.D. thesis. Rutgers University, New Brunswick, New Jersey.

Landenberger, D. E. 1968. Studies of selective feeding in the Pacific starfish *Pisaster* in southern California. Ecology **49**:1062–1075.

Murdoch, W. W. 1969. Switching in general predators: Experiments on predator specificity and stability of prey population. Ecol. Monogr. 39:335–354.

Murdoch, W. W., and J. R. Marks. 1973. Predation by coccinellid beetles: Experiments on switching. Ecology **54**:160–167.

Oatman, E. R. 1972. An ecological study of arthropod populations on apples in northeastern Wisconsin: Population dynamics of mite species on the foliage. Ann. Entomol. Soc. Am. **66**:122–131.

Pearson, O. P. 1966. The prey of carnivores during one cycle of mouse abundance. J. Anim. Ecol. **35**: 217–233.

Turnbull, A. L. 1960. The prey of the spider *Linyphia triangularis* (Glerck) (Araneae, Linyphiidae). Can. J. Zool. **38**:859–873.

Webster, L. 1948. Mites of economic importance in the Pacific Northwest. J. Econ. Entomol. **41**:677–683.

```
OATMAN ER............................
  72 ANN ENT SOC AM      66     122
    SANTOS MA         ECOLOGY      57      390     76
```

FIGURE 30. How a citation is indexed in *Science Citation Index.*

been looking at so far have done. Second, now they are of less importance because the review and text sources have done their job. You no longer need those older articles. You want more recent material that answers questions you have not yet answered, and the references you have collected are going to help you do that.

How Science Citation Index is Constructed

One of the primary indexes you will want to use is the *Science Citation Index (SCI)*. Published since 1961, it is an index to the citations in approximately 2,400 scientific journals. As each issue of the 2,400 journals is published, the staff of SCI takes every citation in the bibliographies of the articles and indexes them according to who wrote the article which includes the citation. For example, in FIGURE 30 we see parts of a journal article: the author, title, and journal title, volume and page, and part of the literature cited section of the article. The *Science Citation Index* staff take that information and convert it to a standardized format illustrated at the bottom of FIGURE 30. This is repeated for every citation in every article, editorial, letter, news item, etc., in each issue of the 2,400 journals covered. Through computer manipulation all of the thousands and thousands of index entries are assembled into one master list -- a citation index (FIGURE 31). In other words, the citation index tells you, for a given time period, who has included whom in their list of cited works.

Citation Index

Perhaps you are now saying, "That's enough about how it is constructed -- how do I use it?" In fact, use is very simple. You know that articles are relevant to your topic (any of those in FIGURE 29). In this case you have primary journal articles on your list. However, in other cases you could have books, government documents, patents, or any other type of publication. Any of these types can be searched in the *Science Citation Index*. By looking up the authors of these articles you can tell who has cited them. For example, G. Caughley's paper is pivotal to your study because as Odum indicated, it disputes the original idea of predator--prey relationships in the Kaibab deer/prey incident. It was written in 1970. You need to know what has been written since 1970 on the same topic. Has Caughley's work been refuted or supplemented? What articles have been written recently on deer/predator relationships? Since it

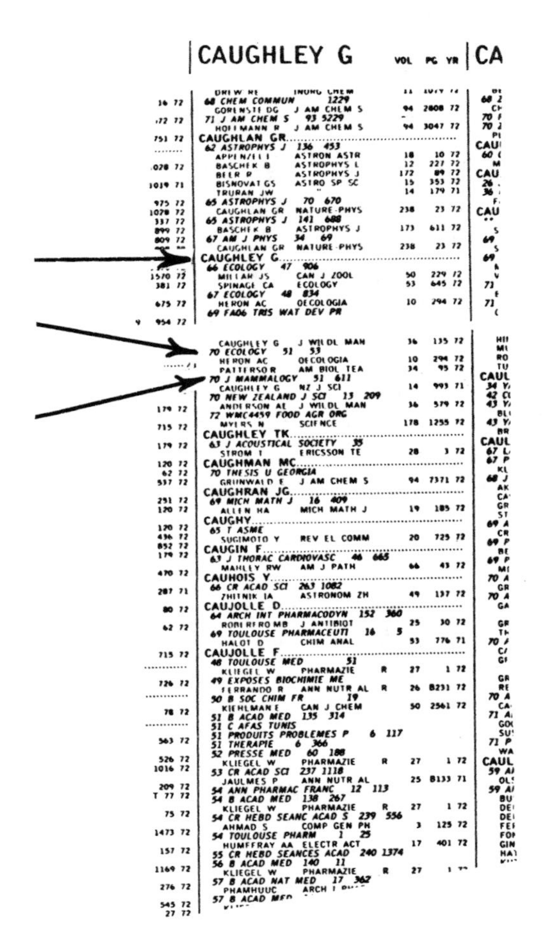
CAUGHLEY G VOL PG YR
68 CHEM COMMUN 1229
GORENSTI DG J AM CHEM S 94 2808 72
71 J AM CHEM S 93 5229
HOFFMANN R J AM CHEM S 94 3047 72
CAUGHLAN GR
62 ASTROPHYS J 136 453
APPENZELLI ASTRON ASTR 18 10 72
BASCHEK B ASTROPHYS L 12 227 72
BEER P ASTROPHYS J 172 89 72
BISNOVAT GS ASTRO SP SC 15 353 72
TRURAN JW " 14 179 71
65 ASTROPHYS J 70 670
CAUGHLAN GR NATURE-PHYS 238 23 72
65 ASTROPHYS J 141 688
BASCHEK B ASTROPHYS J 173 611 72
67 AM J PHYS 34 69
CAUGHLAN GR NATURE-PHYS 238 23 72
CAUGHLEY G
66 ECOLOGY 47 906
MILLAR JS CAN J ZOOL 50 229 72
SPINAGE CA ECOLOGY 53 645 72
67 ECOLOGY 48 834
HERON AC OECOLOGIA 10 294 72
69 FA06 TRIS WAT DEV PR
CAUGHLEY G J WILDL MAN 36 135 72
70 ECOLOGY 51 53
HERON AC OECOLOGIA 10 294 72
PATTERSOR AM BIOL TEA 34 95 72
70 J MAMMALOGY 51 611
CAUGHLEY G NZ J SCI 14 993 71
70 NEW ZEALAND J SCI 13 209
ANDERSON AE J WILDL MAN 36 579 72
72 WMC4459 FOOD AGR ORG
MYERS N SCIENCE 178 1255 72
CAUGHLEY TK
63 J ACOUSTICAL SOCIETY 35
STROM T ERICSSON TE 28 3 72
CAUGHMAN MC
70 THESIS U GEORGIA
GRUNWALD E J AM CHEM S 94 7371 72
CAUGHRAN JG
69 MICH MATH J 16 409
ALLEN HA MICH MATH J 19 185 72
CAUGHY
65 T ASME
SUGIMOTO Y REV EL COMM 20 725 72
CAUGIN F
63 J THORAC CARDIOVASC 46 665
MAHLEY RW AM J PATH 66 43 72
CAUHOIS Y
66 CR ACAD SCI 263 1082
ZHITNIK IA ASTRONOM ZH 49 137 72
CAUJOLLE D
64 ARCH INT PHARMACODYN 152 360
ROBEREROMB J ANTIBIOT 25 30 72
69 TOULOUSE PHARMACEUTI 16 5
HALOT D CHIM ANAL 53 776 71
CAUJOLLE F
48 TOULOUSE MED 51
KLIEGEL W PHARMAZIE R 27 1 72
49 EXPOSES BIOCHIMIE ME
FERRANDO R ANN NUTR AL R 26 B231 72
50 B SOC CHIM FR 19
KIEHLMAN E CAN J CHEM 50 2561 72
51 B ACAD MED 135 314
51 C AFAS TUNIS
51 PRODUITS PROBLEMES P 6 117
51 THERAPIE 6 366
52 PRESSE MED 60 188
KLIEGEL W PHARMAZIE R 27 1 72
53 CR ACAD SCI 237 1118
JAULMES P ANN NUTR AL 25 B133 71
54 ANN PHARMAC FRANC 12 113
54 B ACAD MED 138 267
KLIEGEL W PHARMAZIE R 27 1 72
54 CR HEBD SEANC ACAD S 239 556
AHMAD S COMP GEN PH 3 125 72
54 TOULOUSE PHARM 1 25
HUMFFRAY AA ELECTR ACT 17 401 72
55 CR HEBD SEANCES ACAD 240 1374
56 B ACAD MED 140 11
KLIEGEL W PHARMAZIE R 27
57 B ACAD NAT MED 17 362
PHAMHUUC ARCH
57 B ACAD MED

FIGURE 31. Science Citation Index. 1972. "Citation Index."

is likely that anyone writing on deer/predator relationships will cite Caughley's 1970 paper, you will find those authors under Caughley's name in *SCI*. FIGURE 31 shows you what the 1972 "Citation Index" of *SCI* lists under Caughley's name. Note that Caughley has written several articles which were cited by various authors during 1972. However, you are only interested in his 1970 article in "*Ecology* 51: 53." As FIGURE 31 illustrates, the "Citation Index" indicates that Heron and Patterson have each cited Caughley. You can expect therefore, that Heron's and Patterson's articles have something to do with the subject of Caughley's article -- deer population size and their fluctuation. You might immediately conclude there are two new sources on your topic.

On to the Source Index

On the contrary, you cannot make that assumption. It is important to realize that authors cite other authors for a number of reasons:

1) To give credit for methodology or ideas from one study applied to another,
2) To relate the specifics of one study to that of another which falls within the same theoretical framework,
3) To draw relationships, either similarities or differences between the results of one study and those of another.

Because of this variety of reasons you cannot be sure that the new articles are going to be highly relevant to your topic. For this reason it is impor-tant for you to know something more about the new articles.

To find out more about the new articles you could simply go and look at them. For many reasons this may not be practical, in which case another portion of *SCI* can be used -- the "Source Index." The "Source Index" is an author index to all the articles which were indexed in the "Citation Index," and therefore it is possible to look up Heron's and Patterson's names in the "Source Index." Here, FIGURE 32, you find all the articles published by a particular author during 1972, and therefore you are able to get some idea of author's research interest. (Note: Since only initials are used it is possible two or more authors with the same last name and first initials may be listed together.) In the "Source Index" you will find the title of each article, which should tell you something about the sub-ject of the article.

Tips on the Use of *SCI*

You should search each of your articles through all years of the citation index since each was published. Therefore, Caughley's article should be searched in the "Citation Index" for the years 1970 up through the latest issue of the *SCI*. (*SCI* is published first as quarterlies, with the January--March quarter appearing in May, the April--June quarter in September, and the July--September quarter in November. The fourth quarter is never published separately but rather is part of the annual volumes published in May or June of the following year. Cumulations of *SCI* have also been published, and therefore you may not need to search a long series of annual volumes).

As you search through the "Citation Index," beginning with the year your original article was published, you will discover articles on your topic. You should incorporate them into your list and should also search them in the "Citation Index."

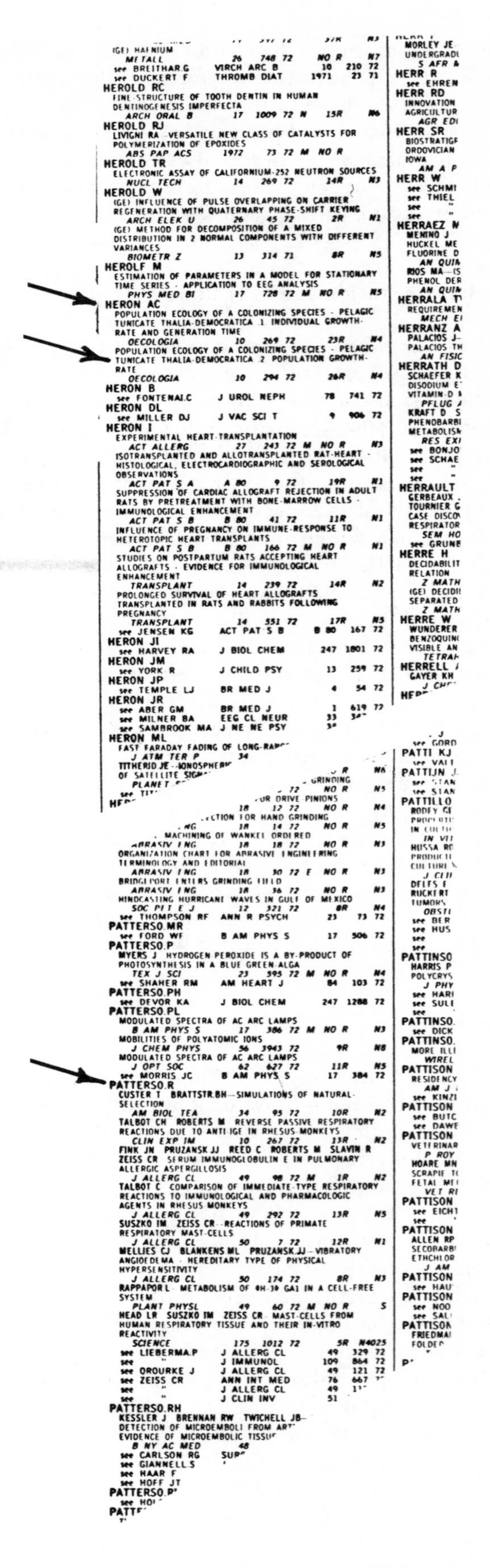
(GE) HAFNIUM
METALL 26 748 72 NO R N7
see BREITHAR.G VIRCH ARC B 10 210 72
see DUCKERT F THROMB DIAT 1971 23 71
HEROLD RC
FINE-STRUCTURE OF TOOTH DENTIN IN HUMAN DENTINOGENESIS IMPERFECTA
ARCH ORAL B 17 1009 72 N 15R N6
HEROLD RJ
LIVIGNI RA VERSATILE NEW CLASS OF CATALYSTS FOR POLYMERIZATION OF EPOXIDES
ABS PAP ACS 1972 73 72 M NO R
HEROLD TR
ELECTRONIC ASSAY OF CALIFORNIUM-252 NEUTRON SOURCES
NUCL TECH 14 269 72 14R N3
HEROLD W
(GE) INFLUENCE OF PULSE OVERLAPPING ON CARRIER REGENERATION WITH QUATERNARY PHASE-SHIFT KEYING
ARCH ELEK U 26 45 72 2R N1
(GE) METHOD FOR DECOMPOSITION OF A MIXED DISTRIBUTION IN 2 NORMAL COMPONENTS WITH DIFFERENT VARIANCES
BIOMETR Z 13 314 71 8R N5
HEROLF M
ESTIMATION OF PARAMETERS IN A MODEL FOR STATIONARY TIME SERIES - APPLICATION TO EEG ANALYSIS
PHYS MED BI 17 728 72 M NO R N5
HERON AC
POPULATION ECOLOGY OF A COLONIZING SPECIES - PELAGIC TUNICATE THALIA-DEMOCRATICA 1 INDIVIDUAL GROWTH-RATE AND GENERATION TIME
OECOLOGIA 10 269 72 23R N4
POPULATION ECOLOGY OF A COLONIZING SPECIES - PELAGIC TUNICATE THALIA-DEMOCRATICA 2 POPULATION GROWTH-RATE
OECOLOGIA 10 294 72 26R N4
HERON B
see FONTENAI.C J UROL NEPH 78 741 72
HERON DL
see MILLER DJ J VAC SCI T 9 906 72
HERON I
EXPERIMENTAL HEART-TRANSPLANTATION
ACT ALLERG 27 243 72 M NO R N3
ISOTRANSPLANTED AND ALLOTRANSPLANTED RAT-HEART - HISTOLOGICAL, ELECTROCARDIOGRAPHIC AND SEROLOGICAL OBSERVATIONS
ACT PAT S A A 80 9 72 19R N1
SUPPRESSION OF CARDIAC ALLOGRAFT REJECTION IN ADULT RATS BY PRETREATMENT WITH BONE-MARROW CELLS - IMMUNOLOGICAL ENHANCEMENT
ACT PAT S B B 80 41 72 11R N1
INFLUENCE OF PREGNANCY ON IMMUNE-RESPONSE TO HETEROTOPIC HEART TRANSPLANTS
ACT PAT S B B 80 166 72 M NO R N1
STUDIES ON POSTPARTUM RATS ACCEPTING HEART ALLOGRAFTS - EVIDENCE FOR IMMUNOLOGICAL ENHANCEMENT
TRANSPLANT 14 239 72 14R N2
PROLONGED SURVIVAL OF HEART ALLOGRAFTS TRANSPLANTED IN RATS AND RABBITS FOLLOWING PREGNANCY
TRANSPLANT 14 551 72 17R N5
see JENSEN KG ACT PAT S B B 80 167 72
HERON JI
see HARVEY RA J BIOL CHEM 247 1801 72
HERON JM
see YORK R J CHILD PSY 13 259 72
HERON JP
see TEMPLE LJ BR MED J 4 54 72
HERON JR
see ABER GM BR MED J 1 619 72
see MILNER BA EEG CL NEUR 33
see SAMBROOK MA J NE NE PSY
HERON ML
FAST FARADAY FADING OF LONG-RA
J ATM TER P 34
TITHERID JE --IONOSPHER
OF SATELLITE SIG
PLANET
18 12 72 NO R N4
FOR HAND GRINDING
18 14 72 NO R N5
MACHINING OF WANKEL ORDERED
ABRASIV ENG 18 18 72 NO R N3
ORGANIZATION CHART FOR ABRASIVE ENGINEERING TERMINOLOGY AND EDITORIAL
ABRASIV ENG 18 30 72 E NO R N3
BRIDGEPORT ENTERS GRINDING FIELD
ABRASIV ENG 18 36 72 NO R N3
HINDCASTING HURRICANE WAVES IN GULF OF MEXICO
SOC PET E J 12 321 72 8R N4
see THOMPSON RF ANN R PSYCH 23 73 72
PATTERSO.MR
see FORD WF B AM PHYS S 17 506 72
PATTERSO.P
MYERS J HYDROGEN PEROXIDE IS A BY-PRODUCT OF PHOTOSYNTHESIS IN A BLUE GREEN-ALGA
TEX J SCI 23 595 72 M NO R N4
see SHAHER RM AM HEART J 84 103 72
PATTERSO.PH
see DEVOR KA J BIOL CHEM 247 1288 72
PATTERSO.PL
MODULATED SPECTRA OF AC ARC LAMPS
B AM PHYS S 17 386 72 M NO R N3
MOBILITIES OF POLYATOMIC IONS
J CHEM PHYS 56 3943 72 9R N8
MODULATED SPECTRA OF AC ARC LAMPS
J OPT SOC 62 627 72 11R N5
see MORRIS JC B AM PHYS S 17 384 72
PATTERSO.R
CUSTER T BRATTSTR.BH--SIMULATIONS OF NATURAL-SELECTION
AM BIOL TEA 34 95 72 10R N2
TALBOT CH ROBERTS M REVERSE PASSIVE RESPIRATORY REACTIONS DUE TO ANTI-IGE IN RHESUS-MONKEYS
CLIN EXP IM 10 267 72 13R N2
FINK JN PRUZANSK JJ REED C ROBERTS M SLAVIN R ZEISS CR SERUM IMMUNOGLOBULIN E IN PULMONARY ALLERGIC ASPERGILLOSIS
J ALLERG CL 49 98 72 M 1R N2
TALBOT C COMPARISON OF IMMEDIATE-TYPE RESPIRATORY REACTIONS TO IMMUNOLOGICAL AND PHARMACOLOGIC AGENTS IN RHESUS-MONKEYS
J ALLERG CL 49 292 72 13R N5
SUSZKO IM ZEISS CR--REACTIONS OF PRIMATE RESPIRATORY MAST-CELLS
J ALLERG CL 50 7 72 12R N1
MELLIES CJ BLANKENS.ML PRUZANSK.JJ--VIBRATORY ANGIOEDEMA - HEREDITARY TYPE OF PHYSICAL HYPERSENSITIVITY
J ALLERG CL 50 174 72 8R N3
RAPPAPOR L METABOLISM OF ΦH-3Φ GA1 IN A CELL-FREE SYSTEM
PLANT PHYSL 49 60 72 M NO R S
HEAD LR SUSZKO IM ZEISS CR MAST-CELLS FROM HUMAN RESPIRATORY TISSUE AND THEIR IN-VITRO REACTIVITY
SCIENCE 175 1012 72 5R N4025
see LIEBERMA.P J ALLERG CL 49 329 72
see " J IMMUNOL 109 864 72
see OROURKE J J ALLERG CL 49 121 72
see ZEISS CR ANN INT MED 76 667
see " J ALLERG CL 49
see " J CLIN INV 51
PATTERSO.RH
KESSLER J BRENNAN RW TWICHELL JB--
DETECTION OF MICROEMBOLI FROM AR
EVIDENCE OF MICROEMBOLIC TISSU
B NY AC MED 48
see CARLSON RG SUP
see GIANNELL.S
see HAAR F
see HOFF JT
PATTERSO.P
see HO
PATT

FIGURE 32. Science Citation Index. 1972. "Source Index."

This process is called "cycling," and it permits you to build a larger more complete bibliography of the most current literature on your topic.

In your work with *SCI* you may eventually come to the point where certain names continue to reappear and a select group of articles always refers to some or all of the articles you have on your list. When this happens you have identified the essential core of literature on your topic, as defined by workers in the field. (Do not be dis-mayed if your search of *SCI* does not result in this closely related set of articles. It does not always happen, and it frequently is more a re-flection of the field of study than of the effec-tiveness of your search). When you finish search-ing *SIC* you are ready to study the collection of articles you have identified. You will want to read them carefully, taking notes as you go. Then you will want to relate their findings to those of the earlier articles and fit them into your outline. In studying the new articles you may discover new authors and papers you had not known about before. You will want to read them, and if they are helpful, check them in the "Citation Index."

Other Author Approaches

To conclude this chapter it should be pointed out that if in your search several authors stand out as the preeminent authors on your topic you should do a thorough author search. First use the "Source Index" of the *SCI* and second use the author in-dexes of the indexing tools we explain in Chapter 6 and list in Appendix IV. For example, when you looked at *Advances in Ecological Research* you noted that Mech wrote on moose–wolf interactions (FIGURE 28). If you search Mech in the *Science Citation Index*, you will discover that he has written quite extensively on wolves, their ecology and behavior (FIGURE 33). You might want, therefore, to know what he has written recently that is related to the question of the effect of wolf predation on moose population size. FIGURE 34 shows you the 1974 *SCI* "Source Index" which lists an article by Mech, and FIGURE 35 shows you the *Biological Abstracts* author index which also lists articles by Mech. We explain the num-bers following the author's name in Chapter 6).

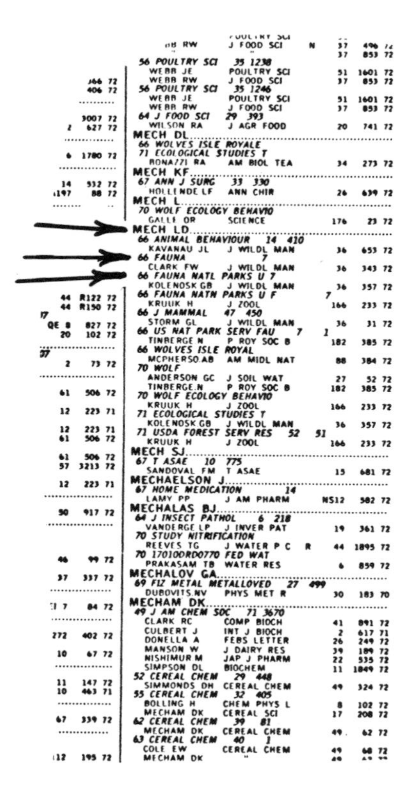

FIGURE 33. Science Citation Index. 1972. "Citation Index."

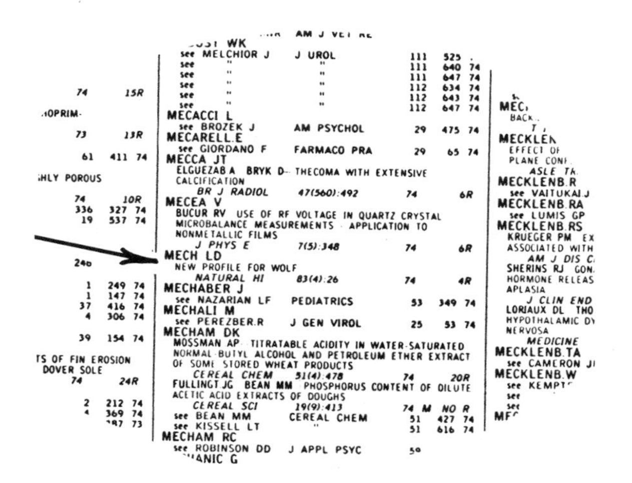

FIGURE 34. Science Citation Index. 1974. "Source Index."

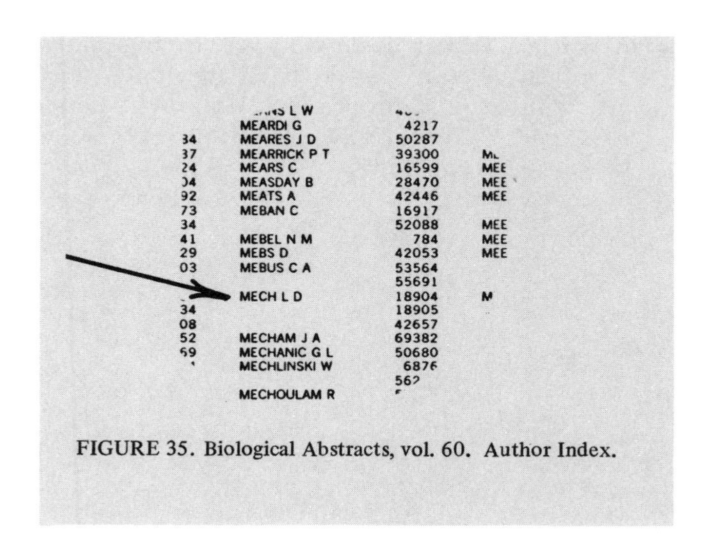

FIGURE 35. Biological Abstracts, vol. 60. Author Index.

Summary

1. There are two approaches for searching periodical literature: the author approach, Chapter 5, and the subject approach, Cahpter 6.
2. The primary tool for an author approach is the *Science Citation Index*.
3. To use *SCI* consult the "Citation Index" under the names of the authors whom you have on your list of relevant articles. This will tell you who has cited the articles.
4. The "Source Index" will tell you what the titles of the citing articles are and therefore help sort out the useful articles.
5. Each article on your list should be searched for all years since it was published up to the present.
6. When certain authors appear to dominate a field of study, a search for articles by them would be appropriate. Use the "Source Index" of *SCI* and the author indexes of such indexing tools as *Biological Abstracts, Chemical Abstracts,* and others. (see Chapter 6)

Apologies

Not every search in *Science Citation Index* is profitable. If your search should turn up little that is useful do not panic. Either ask your librarian for help or go on to the subject approach described in Chapter 6. In doing a search in the various indexing/abstracting services we explain in Chapter 6, authors may appear which you can return to *SCI* and search.

Your library may not have the *Science Citation Index*. That is unfortunate because it is very useful and easy to use. If your library does not have it you may want to consider going to another library. If you do, read Chapter 8, "Using Other Libraries."

6 Biological Abstracts and Other Subject Indexes

Maybe you had to write a paper early in your college career and when it came time to find periodical articles you turned to the *Readers' Guide* because that was the only periodical index you knew. But there are many -- in fact, several hundred -- different indexes to journal articles from which you could have chosen. Each index covers a different group of journals and uses different indexing techniques. Furthermore particular libraries do not have all of the indexes. Therefore when it comes time to use a periodical index for a library or research problem in biology you must ask yourself, "which indexes should I use?"

Readers' Guide covers mostly popular magazines, inappropriate sources of information for a biology paper. No doubt your professor mentioned at some point that you should use *Biological Abstracts,* because it is generally the single best periodical index for the biological literature. However, your library may not have it, or in some cases another index would be more appropriate. At the end of this chapter we will say something about the choice of an appropriate index for a particular search.

The main part of this chapter will deal with four indexes to the biological literature in general: 1) *Biological Abstracts,* and its companion *Bioresearch Index,* 2) *Biological and Agricultural Index,* 3) *Biology Digest,* and 4) "Permuterm Index" of *Science Citation Index.* If it is at all possible you should get to a library that has *Biological Abstracts,* but if that is impossible then you will have to use one of the other three whichever your library has. Accordingly you should read the appropriate section of this chapter on *Biological Abstracts, Biological and Agricultural Index, Biology Digest,* or Permuterm Index of *SCI.*

Biological Abstracts . . . What is it?

Unlike the indexes with which you are probably already familiar (i.e. *Readers' Guide) B.A.* is more than just a periodical index. It might be more properly called an information resources directory. The basic component of *B.A.* is an abstract, which is a summary statement of the content of a document. The abstract tells you in a general way what points the author(s) are trying to make. FIGURE 36 illustrates an abstract from *B.A.* Note that preceding the abstract is all of the bibliographic information you associate with an index entry: author(s), title of the article, title of the journal, volume, pages, and date. In addition two other important pieces of information are given: 1) Following the authors' names is the address of the senior or starred author. This permits users of *B.A.* to communicate directly with authors in whose papers they are interested. 2) At the end of the bibliographic information is a note about the language in which the article and summaries were originally published. This is important because articles in foreign languages are included in *Biological Abstracts.*

B.A. includes as many of the important research publications on biological subjects as is practical. This means not only the major national and international journals are included, but many society publications and regional journals, regardless of nationality or the language in which they are published, and some U.S. government documents are also included. The subjects included range over all fields of the biological sciences: behavioral science, biochemistry, botany, development, ecology, evolution, genetics, microbiology, parasitology, physiology, radiation biology, systematic biology, and virology; and their related subjects: aerospace biology, agriculture, experimental medicine, nutrition, pathology, pharmacology, public health, toxicology, and veterinary science. In all there are approximately 140,000 abstracts a year in *Biological Abstracts!*

B.A. Author Index

Accompanying the abstracts in the semi-monthly issues there are five indexes which provide access to the abstracts, and ultimately the journal articles themselves. These semi-monthly indexes are cumulated every six months at the close of each volume. The author index, FIGURE 35, is an alphabetical list of all the authors, the senior or first author, as well as the junior or second, third, etc. authors of each article abstracted. The number following the author's name is the abstract number for the article which s/he wrote. In our example, Mech, L.D. is the author of articles with abstract numbers 18904, 18905 and 42657. As we mentioned in Chapter 5, if there are a few significant authors writing on your topic it is useful to search specifically for articles by them. The author index of *Biological Abstracts* is one of the places to search. For example, Mech is one of the prominent authors who has studied moose--wolf interactions; so a search of the *B.A.* author index (FIGURE 35) should lead you to abstracts of articles by him (FIGURE 36).

B.A. Subject Index

All of the other indexes in *Biological Abstracts* are some form of subject index, but only one of them is actually called a subject index; it is the one we will discuss next. This index is very dif-- ferent from the *Readers' Guide* and most other indexes you are likely to have used. Because it is so very different we will take a few lines to explain to you how it is constructed. Note the title of Mech's article in FIGURE 36. To create subject index entries to that abstract the producers of *B.A.* have put it into a computer. In addition, the editors take some significant words from the abstract and combine them with the title. The computer is then told to print every significant word from the title and the added words. In addition the editors have told the computer to print a few words from the title that come be--

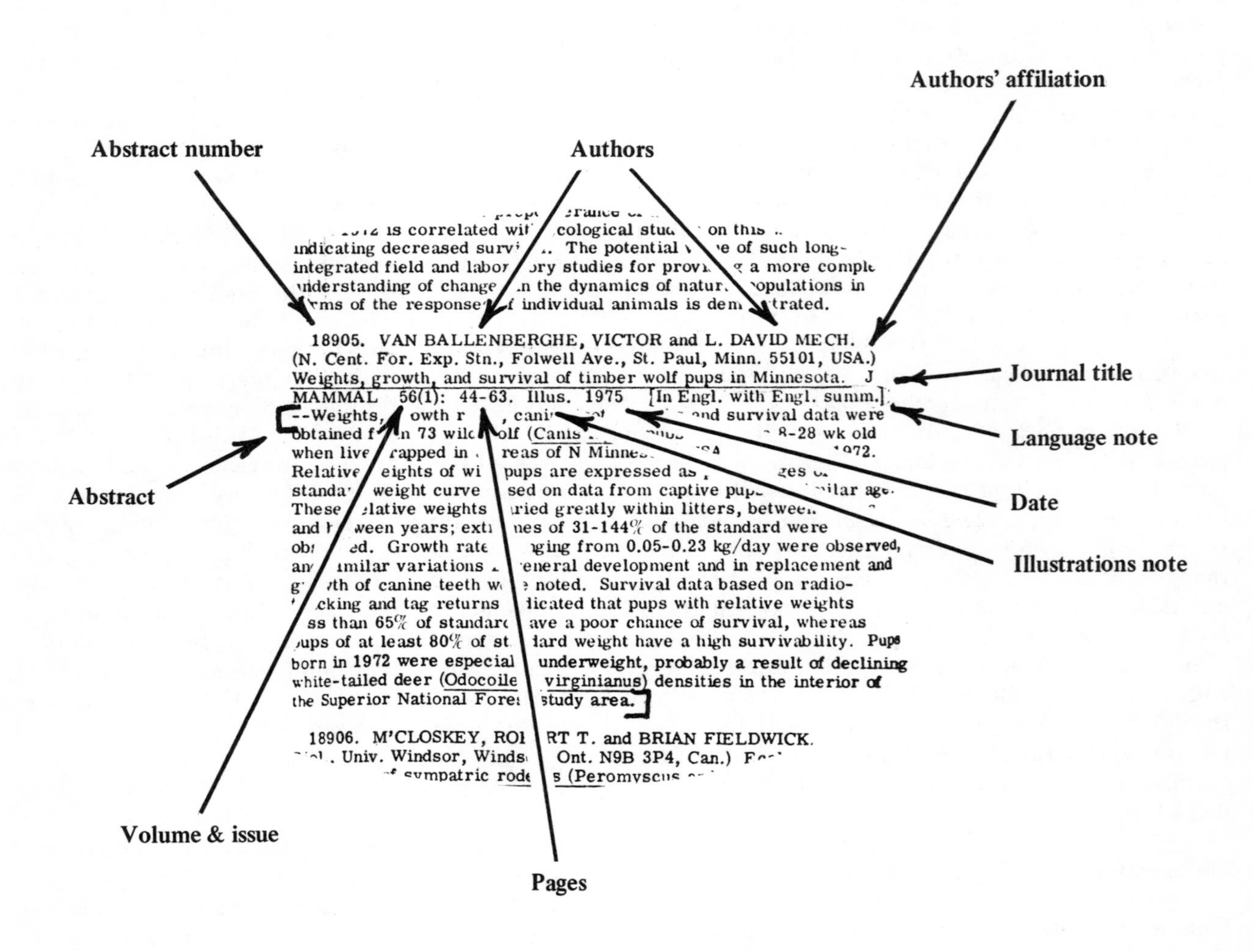
18905. VAN BALLENBERGHE, VICTOR and L. DAVID MECH. (N. Cent. For. Exp. Stn., Folwell Ave., St. Paul, Minn. 55101, USA.) Weights, growth, and survival of timber wolf pups in Minnesota. J MAMMAL 56(1): 44-63. Illus. 1975 [In Engl. with Engl. summ.]

FIGURE 36. Biological Abstracts, vol. 60. Abstracts.

fore and after each significant or key word. The computer, as a result, prints out the title in twelve ways as shown in FIGURE 37. (Note in FIGURE 37 the "/" which is the symbol used to indicate that the next word is the beginning of the title.)

The computer does the same thing for all the other thousands of abstracts, arranging them in one alphabetical list. The result is the subject index to *B.A.*, an example of which is shown in FIGURE 38. (Note that when several lines have the same key word such as "timber," the key word is omitted in all lines after the first. This gives you a readily visable idea of the number of abstracts that have a particular key word in their titles.) It is important to keep these things in mind about the topography of this index:

1) The alphabetical arrangement is based on the key word which is positioned in the middle of the column. There is shading on the left for visual clarity.
2) Each line is a separate title and therefore each line leads to a different abstract.
3) Because the column is limited in width,

9032	PECIES USTILAGINACEAE	TILLETIACEAE PRO MYCELIAL CHARACTER	29899
37972	COPY OF JAPANAUT AND	TILLIGERRY VIRUSES 2 PROPOSED MEMB	69995
12358	PH BIOLOGICAL ACTIVITY	TILLING /MULCH AND MICRO CLIMATE TH	46311
66947	A MODIFICATION OF THE	TILLMAN OLSEN METHOD FOR OBTAINING	5396
1424	GY SEDIMENT/ MULTIPLE	TILLS RADIOMETRIC AGES AND ASSESSME	27137
18604	AN INTERFERON INDUCER	TILORONE DI HYDRO CHLORIDE HERPESVI	41898
34747	TAMYCIN A DAUNOMYCIN	HYDRO CHLORIDE MODIFIED N	54413
34710	ATIVES AND ANALOGS OF	HYDRO CHLORIDE MOUSE CEL	67713
36482	LAMMATORY ACTIONS OF	HYDRO CHLORIDE RAT ADRENA	33438
2643	OCESS OF LATERAL HEAD	TILT /CONSTANT REPRESENTATION OF VI	10059
19052	SFER OF COLOR SPECIFIC	AFTEREFFECTS HUMAN MCCULLOUG	39940
42527	D CHIMPANZEE NEURON/	ANALYSIS OF PLEOMORPHIC VESICLE	38894
59803	ADAPTATION TO OPTICAL	AND DISPLACEMENT HUMAN ORIENT	11158
36331	F VISUAL ADAPTATION TO	AND DISPLACEMENT HUMAN PERCE	11162
57578	ING MUTANTS WITHERED	AND DUMPY PYRIMIDINE METABOLIS	2227
13352	RE/ REEVALUATION OF A	BACK SEAT AS A MEANS OF INCREAS	17513
62479	CARDIAL CONTRACTILITY	ECHO CARDIOGRAPHY/ ALTERATION	18405
66975	HE MAGNITUDES OF THE	ILLUSION AND AFTEREFFECT HUMAN	51748
7411	OLLING WITH HEAD BODY	IN THE MEDIAN PLANE HUMAN/ THE	22899
2636	E EPINEPHRINE HEAD-UP	LOW SALT DIET/ CIRCULATING CATE	50560
65475	PARISON OF 70 DEGREE	LOWER BODY NEGATIVE PRESSURE A	35013
42761	N/ THE EFFECT OF HEAD	ON MERIDIONAL DIFFERENCES IN AC	11159
62646	URINARY EXCRETION AND	TABLE TOLERANCE IN ATHLETES AN	17516
65443	ROGENIC HYPERACTIVITY	TEST/ VASCULAR REACTIVITY TO AN	12567
156	WITH THE BJORK SHILEY	TILTING DISC VALVE PROSTHESIS CLINIC	53130
48322	WITH THE BJORK SHILEY	DISC VALVE PROSTHESIS HUMAN	53131
44017	ROLLING AS A RESULT OF	HUMAN/ A COMPARISON BETWE	17082
13269	THE NETHERLANDS AVES	TIMALIINAE /POST JUVENILE MOLT OF TH	41963
	IN THE SHOREA-ROBUSTA	TIMBER DURING PATHOGENESIS BY IRPEX	22178
47381	INGDOM APPLE BRUISING	FIBERBOARD/ COMPARISONS OF	43214
13538	BANCE ASSOCIATED WITH	HARVESTING IN SOUTHWESTERN	48879
60936	PROCESS WATER IN THE	INDUSTRY CORYNEBACTERIUM BA	45972
24500	7 FOREST ECOLOGY AND	MANAGEMENT RE FORESTATION	8199
6927	RA-FUMIFERANA VIRUSES	PEST FUNGI BACTERIA MICRO SP	60038
24110	RINATING ENZYME FROM	RATTLESNAKE CROTALUS-HORRID	21622
23007	PROBABLE ROLE OF THE	RATTLESNAKE CROTALUS-HORRID	27157
36405	LUS-HORRIDUS-HORRIDUS	RATTLESNAKE HEMATOMA STAPH	17340
15968	ECTED SHOREA-ROBUSTA	UTILIZATION/ ESTIMATION OF CE	22179
48134	MANS/ ECOLOGY OF THE	WOLF IN NORTHEASTERN MINNES	24683
	OWTH AND SURVIVAL OF	WOLF PUPS IN MINNESOTA USA	18905
58285	BSERVATIONS ABOVE THE	TIMBERLINE AT SAFED KOH EAST AFGHA	59883
53413	ECOB HUMAN REACTION	TIME /A CONTENT ADDRESSABLE MODEL	22532

FIGURE 38. Biological Abstracts, vol. 60. Subject Index.

1)	MAL NUTRITION/ WEIGHTS	GROWTH AND SURVIVAL OF TIMBER WOLF P	18905
2)	ADIO TRACKING TAGGING	MAL NUTRITION/ WEIGHTS GROWTH AND	18905
3)	OF TIMBER WOLF PUPS IN	MINNESOTA USA ODOCOILEUS-VIRGINIANUS	18905
4)	TRACKING TAGGING MAL	NUTRITION/ WEIGHTS GROWTH AND SUR	18905
5)	SURVIVAL OF TIMBER WOLF	PUPS IN MINNESOTA USA ODOCOILEUS-VIR	18905
6)	IRGINIAUS CANIS-LUPUS	RADIO TRACKING TAGGING MAL NUTRITIO	18905
7)	ION/ WEIGHTS GROWTH AND	SURVIVAL OF TIMBER WOLF PUPS IN MINN	18905
8)	-LUPUS RADIO TRACKING	TAGGING MAL NUTRITION/ WEIGHTS GRO	18905
9)	NUS CANIS-LUPUS RADIO	TRACKING TAGGING MAL NUTRITION/ WEI	18905
10)	OLF PUPS IN MINNESOTA	USA ODOCOILEUS-VIRGINIANUS CANIS-LUP	18905
11)	TAGGING MAL NUTRITION/	WEIGHTS GROWTH AND SURVIVAL OF TIMBE	18905
12)	AND SURVIVAL OF TIMBER	WOLF PUPS IN MINNESOTA USA ODOCOILEU	18905

FIGURE 37. Title of Mech's article, with enrichment words, as permuted by the computer in twelve locations within the alphabetical sequence in Biological Abstracts subject index.
Title: Weights growth and survival of timber wolf pups in Minnesota.
Enrichment words: USA Odocoileus virginianus Canis lupus Radio Tracking Tagging Malnutrition

and because the computer is unable to distinguish syllables the lines in the index are only part of the whole title and may not make sense because of incompleteness. Those odd letter combinations at the be-- ginning and end of many lines are word fragments from the preceding or suc-- ceeding words of the title.

"But what does all of this mean for my search?" you are asking. Quite simply it means that you must know a fair amount about your subject, particularly the terms which are most frequently used in association with it. This is where the list of terms you compiled in FIGURE 29 becomes useful. Since these are terms which are frequently associated with discussions of your topic, these are the terms you should use in searching the in-- dex. In using these terms you will no doubt need to experiment a little. For example, broad terms like "Population" are likely to produce too long a list of articles to be able to search efficiently, while other terms may turn out to be unused by authors and therefore do not show up in the index. After you have tried several possibilities you will discover that certain terms are better than others. In FIGURE 39 examples of what you would find in *Biological Abstracts* under "Pred-- ator" and "Deer," are shown.

There are a large number of entires under each of these terms, and it is therefore necessary to scan the titles and try to *eliminate* those that are clearly irrelevant to your search. It is unlikely that you will be able to select the most useful articles directly from the index. Usually you need to copy down a list of abstract numbers and then check each abstract for the full title and summary. Your search can be expedited if you note the titles carefully while scanning them. Watch particularly for familiar terms and word combinations. The arrangement of titles with the same key word alphabetical ac-- cording to the succeeding words is helpful for these word combinations (FIGURE 39). While this process is time--consuming it generally re-- wards the diligent with a few highly useful art-- icles.

As you use *B.A.*'s subject indexes you will, no doubt, have questions about something which we cannot anticipate and explain here. You should consult with your librarian or professor for help when that happens.

Once you have listed the numbers of articles you cannot eliminate, you search for and read the abstracts and then choose those that are most relevant. When you have identified appro-- priate articles on your topic, you will want to lo--

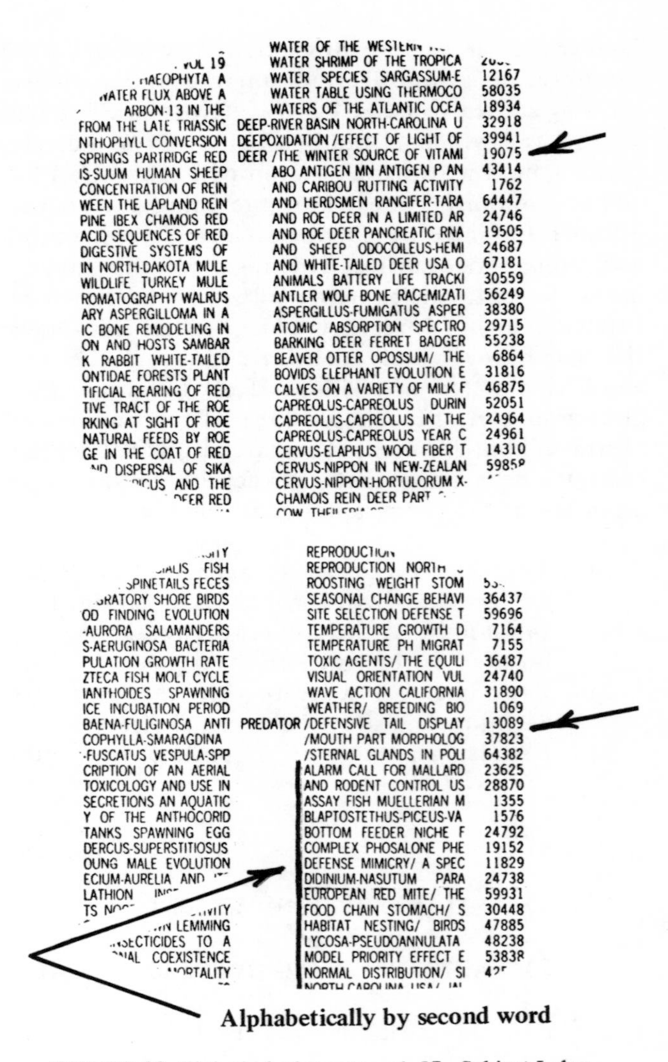

FIGURE 39. Biological Abstracts, vol. 57. Subject Index.

cate them in the library. To do this you will need to consult your library's periodicals list or serials record just as you did after identifying articles in the *Science Citation Index.*

B.A. Organism Indexes

When you are searching for material on specific organisms, the names of the organisms are the best index terms to use. *Biological Abstracts* has acknowledged that fact through the creation of two indexes which give organism--name access to the abstracts. Once again that list of terms in FIGURE 29 is very useful. In it are the com-- mon and scientific names for the organisms in which you are interested. In order to do a thor-- ough search for information on an organism you should search both the common name in the sub--

ject index, as we did for "Deer" in FIGURE 39, and the scientific name, which in the case of deer, is *Odocoileus virginianus.* The genus--species names of organisms are included in the subject index until January, 1974. After that date they are listed in a separate index, the "Generic Index." FIGURE 40 shows you a section of the "Generic Index" and in the first column is the genus--species name for the moose, *Alces alces.* The second col--umn is an abbreviated statement about the subject of the article. This is one of more than 600 dif--ferent subjects which are the same headings under which abstracts are grouped in the "Abstracts" section of *B.A.* While the meaning of most abbre--viations are obvious, as in the case with "Ecology--Animal," others are not, in which case you should consult the master list of headings included with each "Generic Index."

The title of the abstract, in FIGURE 41, which the "Generic Index" leads you to, is a good ex--ample of why a name approach is a useful way to get at the material on your subject. The title con--tains only three words that are on your list (FIG--URE 29): "wolf," "moose," and "population." It would be a very time--consuming process to have found that title under "population." That leaves only the names, "wolf" and "moose," both common and scientific, as effective ways to get at the article. And what an article it is. It is exactly what you want! The interesting fact to contem--plate is that none of the sources we have thus far used lead to the Zarnoch article (FIGURE 41). Furthermore, the articles we found elsewhere (i.e. *Science Citation Index*) have not turned up in our search of *Biological Abstracts.* This ex--ample shows how the use of a number of different tools is entirely appropriate, even necessary, for finding material on a given subject.

We mentioned above that there are two dif--ferent organism indexes; and we are now going to look at the second one, called the "Biosystematic Index." It should be used when you are seeking information that is multi--generic, i.e. when you

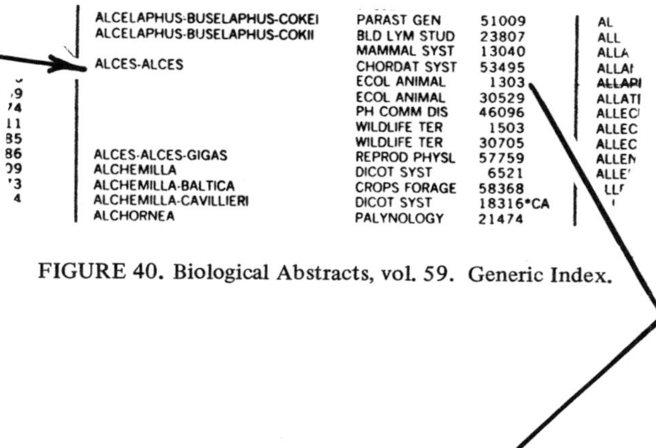

ALCELAPHUS-BUSELAPHUS-COKEI	PARAST GEN	51009
ALCELAPHUS-BUSELAPHUS-COKII	BLD LYM STUD	23807
	MAMMAL SYST	13040
ALCES-ALCES	CHORDAT SYST	53495
	ECOL ANIMAL	1303
	ECOL ANIMAL	30529
	PH COMM DIS	46096
	WILDLIFE TER	1503
	WILDLIFE TER	30705
ALCES-ALCES-GIGAS	REPROD PHYSL	57759
ALCHEMILLA	DICOT SYST	6521
ALCHEMILLA-BALTICA	CROPS FORAGE	58368
ALCHEMILLA-CAVILLIERI	DICOT SYST	18316*CA
ALCHORNEA	PALYNOLOGY	21474

FIGURE 40. Biological Abstracts, vol. 59. Generic Index.

... prey. The ...therms in the mainland ... C. enydris is an opportunistic ...y on ectotherms in spite of the specializa-...or the detection and procurement of endothermic

1303. ZARNOCH, STANLEY J.* and BRIAN J. TURNER. (Div. For. Wildl. Resour., Va. Polytech. Inst. State Univ., Blacksburg, Va., USA.) A computer simulation model for the study of wolf-moose population dynamics. J ENVIRON SYST 4(1): 39-51. Illus. 1974. [In Engl. with Engl. summ.]--A computer simulation model was developed of a wolf-moose (Canis lupus-Alces alces) system. The program is formulated in APL (A Programming Language) which is an interactive language between user and computer. This language is ideal for use as a teaching aid because the student is continuously in contact with the simulation problem. Parameters in the program which may be externally manipulated by the user consist of initial age and sex distributions of the moose and wolf populations, vegetation supply (her-baceous and browse) available for the moose, moose hunting season, length of the simulation run and output options. The other parameters in the model are reproductive rates (age specific) and mortality rates (age and sex specific) for moose and wolves, wolf predation rate upon moose (age specific), vegetation consumption by moose and human hunting success on moose (age and sex specific). These can be altered by internal programming which may be performed with minimal com-puter training.

1304. JOHNSON, MARTIN W. (Scripps Inst. Oceanogr., Univ. Calif., San Diego, La Jolla, Calif. 92037, USA.) On the dispersal of lobster larvae into the East Pacific barrier (Decapoda, Palinuridea). U S NATL MAR FISH SERV FISH BULL 72(3): 639-64... [In Engl. with Engl. summ.]--The seaw... of lobsters occurring along th... tropical Pacific ...

FIGURE 41. Abstract of Zarnoch article IN Biological Abstracts, vol. 59.

want information on several genera. You can and should look up each genus if the number of genera is small. However, for searches of a number of genera, or where you want information on a Family or Order, the "Biosystematic Index" is generally useful.

If you refer back to your master outline and list of terms (FIGURE 29), you will note that you included the family name Cervidae on that list. To use the "Biosystematic Index" you must place your organism within the structure of the index. The arrangement is by large organism groups which are subdivided by subgroups. The basis for the sequence is from simple to complex. Below is the outline of that structure:

Organisms -- General
Microorganisms -- General
Viruses
Bacteria
Plantae Cryptogamae
Algae
Fungi
Lichens
Bryophyta
Tracheophyta
Pteredophyta
Spermatophyta
Gymnospermae
Angiospermae
Monocot
Dicot
Paleobiology
Paleobotany
Paleozoology
Animalia
Protozoa
Invertebrata
Metazoa
Mesozoa
Porifera
Cnidaria
Ctenophora
Helminthes
Platyhelminthes
Rhynchocoela
Acanthocephala
Aschelminthes
Entoprocta
Phoronidea
Ectoprocta
Brachiopoda
Mollusca
Sipunculoidea
Annelida
Echiuroidea
Linguatulida
Tandigrada
Onychophora
Arthropoda
Crustacea
Myriapoda
Insects
Chelicerata
Chastognatha
Hemichordata
Pogonophora
Echisnodermata
Chordata
Prochordata
Vertebrata
Pisces
Amphibia
Reptilia
Aves
Mammalia

Each of these units is subdivided. First, articles on the group as a whole are listed, then articles on subgroups, either Order or Family, are listed in alphabetical order. FIGURE 42 gives you a composite of portions of several pages from the "Biosystematic Index" which refer to articles on the deer or Cervidae. When you find the appropriate groups you then must decode subject category abbreviations which are the same as those in the

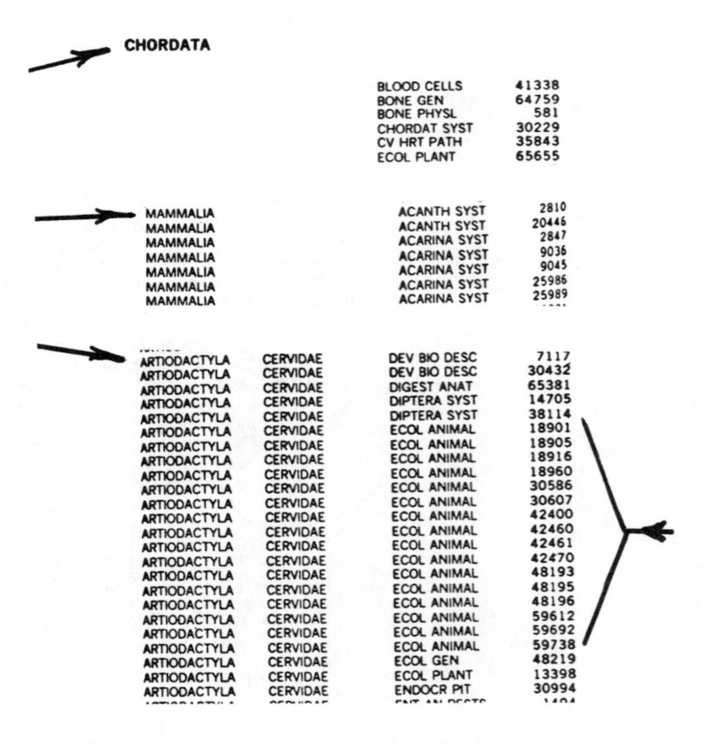

CHORDATA

		BLOOD CELLS	41338
		BONE GEN	64759
		BONE PHYSL	581
		CHORDAT SYST	30229
		CV HRT PATH	35843
		ECOL PLANT	65655
MAMMALIA		ACANTH SYST	2810
MAMMALIA		ACANTH SYST	20446
MAMMALIA		ACARINA SYST	2847
MAMMALIA		ACARINA SYST	9036
MAMMALIA		ACARINA SYST	9045
MAMMALIA		ACARINA SYST	25986
MAMMALIA		ACARINA SYST	25989
ARTIODACTYLA	CERVIDAE	DEV BIO DESC	7117
ARTIODACTYLA	CERVIDAE	DEV BIO DESC	30432
ARTIODACTYLA	CERVIDAE	DIGEST ANAT	65381
ARTIODACTYLA	CERVIDAE	DIPTERA SYST	14705
ARTIODACTYLA	CERVIDAE	DIPTERA SYST	38114
ARTIODACTYLA	CERVIDAE	ECOL ANIMAL	18901
ARTIODACTYLA	CERVIDAE	ECOL ANIMAL	18905
ARTIODACTYLA	CERVIDAE	ECOL ANIMAL	18916
ARTIODACTYLA	CERVIDAE	ECOL ANIMAL	18960
ARTIODACTYLA	CERVIDAE	ECOL ANIMAL	30586
ARTIODACTYLA	CERVIDAE	ECOL ANIMAL	30607
ARTIODACTYLA	CERVIDAE	ECOL ANIMAL	42400
ARTIODACTYLA	CERVIDAE	ECOL ANIMAL	42460
ARTIODACTYLA	CERVIDAE	ECOL ANIMAL	42461
ARTIODACTYLA	CERVIDAE	ECOL ANIMAL	42470
ARTIODACTYLA	CERVIDAE	ECOL ANIMAL	48193
ARTIODACTYLA	CERVIDAE	ECOL ANIMAL	48195
ARTIODACTYLA	CERVIDAE	ECOL ANIMAL	48196
ARTIODACTYLA	CERVIDAE	ECOL ANIMAL	59612
ARTIODACTYLA	CERVIDAE	ECOL ANIMAL	59692
ARTIODACTYLA	CERVIDAE	ECOL ANIMAL	59738
ARTIODACTYLA	CERVIDAE	ECOL GEN	48219
ARTIODACTYLA	CERVIDAE	ECOL PLANT	13398
ARTIODACTYLA	CERVIDAE	ENDOCR PIT	30994

FIGURE 42. Biological Abstracts, vol. 60. Biosystematic Index.

"Generic Index."

This index appears to be difficult because of the large number of categories listed above. It is, in fact, very easy. Just remember that the categories are roughly in an ascending evolutionary arrange-ment for large groups and then alphabetically by Order and/or Family.

By this time you may be wondering how many other indexes there are, and whether you need to use them all? The answer to the first question is that there is only one more index. And the answer to the second is an emphatic *no*! Aren't you re-lieved? But it is important that you know about the existence of each of these indexes and be able to choose the proper ones for each search.

B.A. Cross Index

The fifth and final index to the abstracts in *B.A.* is the "CROSS Index," which is an acronym for "Computer Rearrangement of Subject Speciali-ties." When we first introduced *B.A.* we said the abstracts were grouped together by subject. It is feasible to print an abstract under only one sub-ject, although because of its nature it could appro-priately be listed under any of several different subjects. For example, the Mech article, abstract 18905, in FIGURE 36 is listed under "Ecology" in the abstracts section but in the "CROSS Index" is listed under "Reproductive System – General," as well as "Ecology." This index therefore gives you a complete list of all articles on a general topic, irrespective of where the abstracts actually appeared.

Most uses of the "CROSS Index" are based on doing what its name says: crossing it with other parts of itself or another index. For example if your topic were the "physiological effects on wolves of declining deer populations," you could locate wolves in the "Subject Index" or "Generic Index," and then check particular numbers against the appropriate subject categories in the "CROSS Index." FIGURE 43 shows a composite of those sections of the various *B.A.* indexes which, when seen together, tell you that you have located an article on the "Reproductive System of Wolves as related to Animal Ecology."

This example is probably a little unusual for the use of the "CROSS Index" since the topic has other very concrete subject access points (i.e. taxonomy). However, if your topic had been

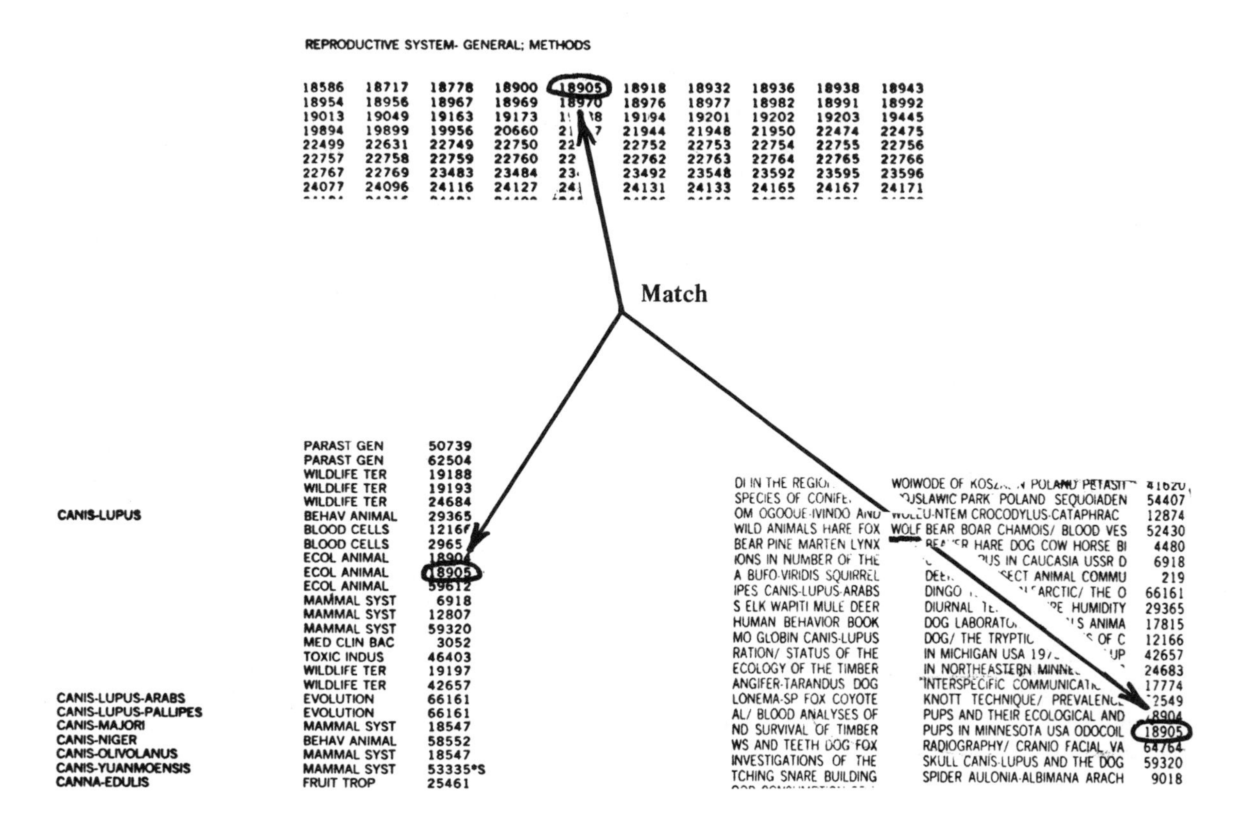

FIGURE 43. Biological Abstracts. Use of CROSS Index in conjunction with Generic or Subject Index. vol.60.

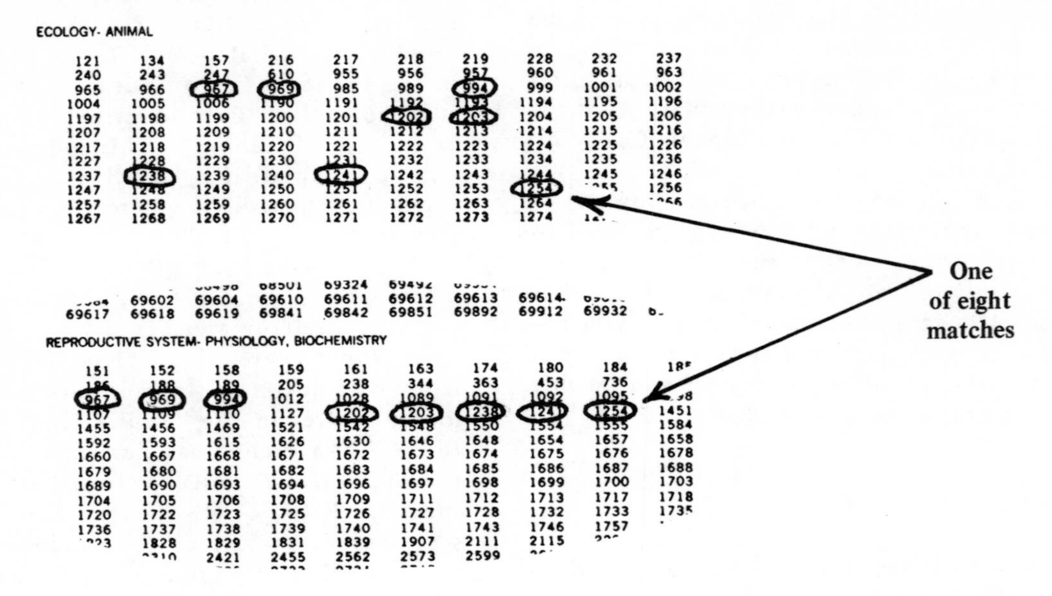

FIGURE 44. Use of the CROSS Index of Biological Abstracts. Vol. 60.

something like "the effects of ecological conditions on reproductive physiology of animals," you would be hard pressed to find effective handles for subject access through any of the other indexes. Remember that the "Subject Index" cannot effectively be used unless you have specific terms that are likely to be used in the titles of research articles. When general terms like "ecological" or "reproductive physiology" are used, long lists of articles are generally found, lists which are too long to search effectively. But as FIGURE 44 shows, it can be done quickly using the "CROSS Index" and comparing the numbers under "Ecology" and "Reproductive System -- Physiology" to find those listed under both.

Parting Thoughts

Biological Abstracts is a difficult tool to use and demands a fair amount of time. However it is the best tool available for searching the world's biological literature. As the well known song goes, "I beg your pardon. I never promised you a rose garden." But we almost certainly can guarantee that with some careful and imaginative use *B.A.* will lead you to useful material. If it does not, please do not ask for "your money back;" instead, ask your reference librarian for help and together "you shall overcome."

Bioresearch Index

Bioresearch Index (Bio. I.) is a companion publication to *Biological Abstracts* and covers the "other half" of the world's biological literature. *Bio. I.* indexes individual articles in collected volumes, symposium proceedings, and review publications; U.S. government documents; abstracts of papers which were presented at meetings; abstracts of documents on biology which appear in other abstracting services; and some publications of less importance, like foreign language applied biology journals or very minor local or regional publications. Generally speaking these are items which are difficult to abstract or which are themselves abstracts.

Fortunately, all the indexes in *Bio. I.* are identical to those in *Biological Abstracts.* Therefore, once you have mastered those in *B.A.* it will be easy to step into using *Bio. I.* The only difference between the two is the way in which the publications are listed. In *Bio. I.* only the basic bibliographic information is provided. Each book, journal issue, or other separate item is listed between rows of stars (FIGURE 45) giving title, volume, issue, number, and date, with the individual chapters, articles or parts listed below. For each the author, title, and paging are given and a number is assigned which is used in the indexes. Once you locate something in one of the indexes you would

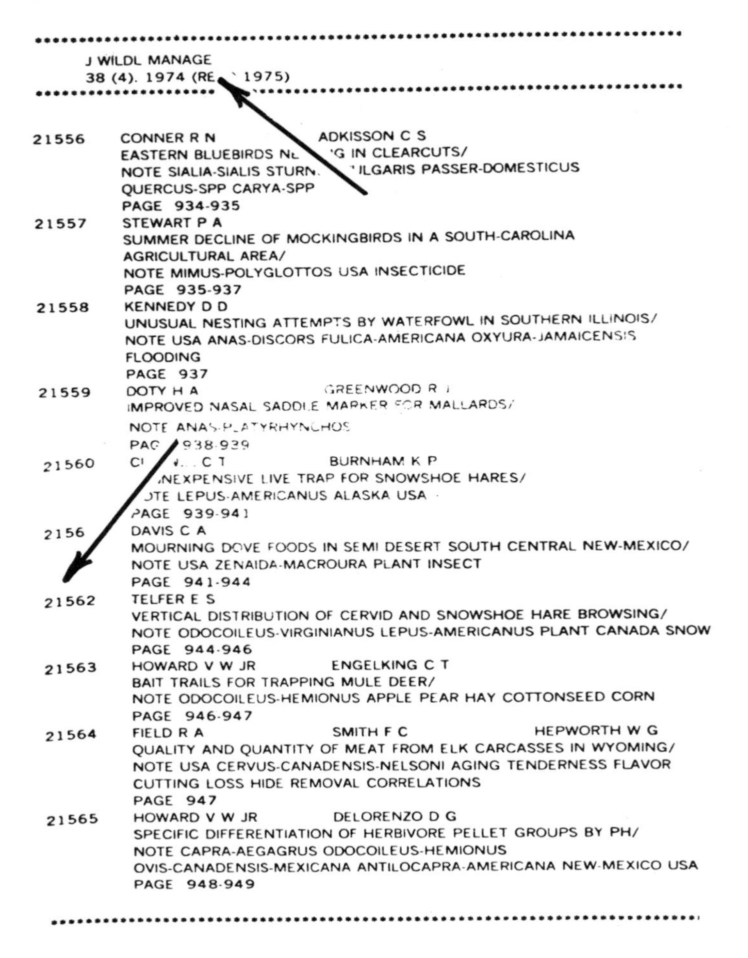

J WILDL MANAGE
38 (4). 1974 (RE ` 1975)

21556 CONNER R N ADKISSON C S
EASTERN BLUEBIRDS N 'G IN CLEARCUTS/
NOTE SIALIA-SIALIS STURN 'ILGARIS PASSER-DOMESTICUS
QUERCUS-SPP CARYA-SPP
PAGE 934-935

21557 STEWART P A
SUMMER DECLINE OF MOCKINGBIRDS IN A SOUTH-CAROLINA AGRICULTURAL AREA/
NOTE MIMUS-POLYGLOTTOS USA INSECTICIDE
PAGE 935-937

21558 KENNEDY D D
UNUSUAL NESTING ATTEMPTS BY WATERFOWL IN SOUTHERN ILLINOIS/
NOTE USA ANAS-DISCORS FULICA-AMERICANA OXYURA-JAMAICENSIS
FLOODING
PAGE 937

21559 DOTY H A GREENWOOD R
IMPROVED NASAL SADDLE MARKER FOR MALLARDS/
NOTE ANAS-PLATYRHYNCHOS
PAGE 938-939

21560 C C T BURNHAM K P
INEXPENSIVE LIVE TRAP FOR SNOWSHOE HARES/
TE LEPUS-AMERICANUS ALASKA USA
PAGE 939-941

2156 DAVIS C A
MOURNING DOVE FOODS IN SEMI DESERT SOUTH CENTRAL NEW-MEXICO/
NOTE USA ZENAIDA-MACROURA PLANT INSECT
PAGE 941-944

21562 TELFER E S
VERTICAL DISTRIBUTION OF CERVID AND SNOWSHOE HARE BROWSING/
NOTE ODOCOILEUS-VIRGINIANUS LEPUS-AMERICANUS PLANT CANADA SNOW
PAGE 944-946

21563 HOWARD V W JR ENGELKING C T
BAIT TRAILS FOR TRAPPING MULE DEER/
NOTE ODOCOILEUS-HEMIONUS APPLE PEAR HAY COTTONSEED CORN
PAGE 946-947

21564 FIELD R A SMITH F C HEPWORTH W G
QUALITY AND QUANTITY OF MEAT FROM ELK CARCASSES IN WYOMING/
NOTE USA CERVUS-CANADENSIS-NELSONI AGING TENDERNESS FLAVOR
CUTTING LOSS HIDE REMOVAL CORRELATIONS
PAGE 947

21565 HOWARD V W JR DELORENZO D G
SPECIFIC DIFFERENTIATION OF HERBIVORE PELLET GROUPS BY PH/
NOTE CAPRA-AEGAGRUS ODOCOILEUS-HEMIONUS
OVIS-CANADENSIS-MEXICANA ANTILOCAPRA-AMERICANA NEW-MEXICO USA
PAGE 948-949

FIGURE 45. Bioresearch Index, vol. 12.

go to the appropriate number in the listing. If the item appears relevant and you want to see it, you will need both the information between the rows of stars above your number *and* the information opposite the number.

Before leaving this chapter you should read the brief section on choosing a periodical index at the end of this chapter.

Biological and Agricultural Index (B.&A.I.)

B.& A.I. is a companion to *Readers' Guide.* It is published by the same company and has the same general format as *R.G.* *B.&A.I.* covers some 150 journals in the biological and agricul-tural sciences. This of course is only a fraction of the literature as compared, say, with *Biological Abstracts* and *Bioresearch Index,* which index several thousand titles, and *Science Citation Index,* which covers approximately 2400 titles.

Subject entries in the *B.&A.I.* consist of con-cepts and ideas, chemical compounds, organism names, and other concrete objects and biological processes. The indexing is standardized in the sense that all articles on a particular subject are listed under a particular term and are not scat-tered under synonyms or whatever terms the au-thor happened to use in the title.

For your topic on deer predators' roles in controlling deer population size the terms you listed in FIGURE 29 are a good place to start your search in the *Biological and Agricultural Index.* A quick check of the index under "preda-tion" (FIGURE 41) indicates that "Predation (Zoology)" is the correct entry. In the example the third article looks like it will be useful. Note that the heading has been subdivided by major groups of animals, and under each are listed articles on predator activities within the group. FIGURE 46 also shows a cross-reference from a heading that is not used, "Predatory Animals," to the actual heading, "Animals, Predatory."

PRECOOLING of fruit. See Fruit, Cooling of
PREDATION (zoology)
Adaptive change in plankton morphology in response to size-selective predation; a new hypothesis of cyclomorphosis. S. I. Dodson. bibl il Limn & Ocean 19:721-9 S '74
Perspectives on predator management. J. Johnson and F. R. Gartner. bibl il J Range Mgt 28:18-21 Ja '75
Relationships between predator removal and white-tailed deer net productivity. S. L. Beasom. bibl il J Wildlife Mgt 38:854-9 O '74
Selectivity of predator control techniques in south Texas. S. L. Beasom. bibl il J Wildlife Mgt 38:837-44 O '74

Arthropods

A study of arthropod predation of pieris rapae L. using serological and exclusion techniques. J. W. Ashby. bibl il J App Ecol 11:419-25 Ag '74

Birds

Consumption of jack-pine budworm, choristoneura pinus Freeman, by the eastern chipping sparrow, spizella passerina (Bechstein) G. A. Simmons and N. F. Sloan. bibl il Can J Zool 52:817-21 Jl '74
Differential predation on gerbils (meriones) by the little owl, athene brahma. D. M. Lay. bibl il J Mammal 55:608-14 Ag '74
Eagles and sheep; a viewpoint. E. G. Bolen. bibl il J Range Mgt 28:11-17 Ja '75

Fishes

The role of predation in the evolution of social behavior of natural populations of the

Invertebrates

On predation of pelagic larvae and early juveniles of marine bottom invertebrates by adult benthic invertebrates and their passing alive through their predators. S. A. Mileikovsky. bibl Marine Biol 26:303-11 S 16 '74

Mathematical models

Mutual interference between parasites or predators and its effect on searching efficiency. J. R. Beddington. bibl il J Animal Ecol 44:331-40 F '75
PREDATORY animals. See Animals, Predatory
PREDNISOLONE
See also
Fluoroprednisolone
PREGEIJEREN
Pregeijeren, Hauptkomponente des ätherischen Wurzelöls von Ruta graveolens. K. H.

FIGURE 46. Biological and Agricultural Index. July, 1975.

In searching for the names of organisms in *B.&A.I.* it is not possible to predict whether it will be listed under the common name or the scientific name. For example the deer mouse is listed under *"Peromyscus,"* the generic name, while deer are listed under "Deer." Since your topic includes all types of North American deer, you should search under the heading, "Elk," but "Venison" can no doubt be ignored since that refers to deer meat. Like the subject "Predation," "Deer" has subdivisions (FIGURE 47). These are in two groups. The first group is subject treatment of deer, for example "Ecology." A second group, separately alphabetized, is the geographical sub-divisions.

Before leaving this chapter you should read the brief section on choosing a periodical index at the end of this chapter.

Biology Digest

Biology Digest is a new entry in the field of indexing the literature of the biological sciences. It is intended primarily for high school libraries, and for small public and academic libraries who cannot afford the more expensive *Biological Abstracts* or even *Biological and Agriculture Index.* It has the advantage, for the user like yourself, of having summaries (abstracts) of the articles indexed while *B.&A.I.* does not. On the other hand its coverage is limited to approximately 220 periodicals, considerably fewer than *Biological Abstracts.* The coverage of journals is different from either *B.A.* or *B.&A.I.* *Biology Digest* includes articles of biological import from general popular journals (i.e. *Time*), from general science periodicals (i.e. *Science*), conservation and pollution periodicals, as well as the basic research journals of biology.

The index (FIGURE 48) is computer generated.

DEEP-sea manganese nodules. See Manganese nodules
DEEP sea mining. See Ocean mining
DEEP sea photography. See Photography, Submarine
DEER
See also
Elk
Venison
also Deer as final subhead under special subjects, e.g. Adrenal glands—Deer

Diseases and pests

Cestodes of white-tailed deer (odocoileus virginianus) in the southeastern United States. A. K. Prestwood. J Parasitology 57:1292 D '71
See also
Catarrhal fever, Malignant
Trichostrongylus dosteri

Ecology

Association measures applied to moose and white-tailed deer distributions. E. Telfer. il J Mammal 53:196-7 F '72
Ecological framework for deer management. H. L. Short. bibliog il J For 70:200-3 Ap '72
Effect of forest manipulation on deer habitat in giant sequoia. G. Lawrence and H. Biswell. bibliog il J Wildlife Mgt 36:595-605 Ap '72
Mule deer fecal group counts related to site factors on winter range. A. E. Anderson and others. bibliog il J Range Mgt 25:66-8 Ja '72

Studies on the biology of an antlered female mule deer. G. W. Mierau. bibliog J Mammal 53:403-4 My '72

Colorado

Adrenal weight in a Colorado mule deer population. A. E. Anderson and others. bibliog il J Wildlife Mgt 35:689-97 O '71
Indices of carcass fat in a Colorado mule deer population. A. E. Anderson and others. bibliog il J Wildlife Mgt 36:579-94 Ap '72

Denmark

Factors affecting growth and body size of roe deer. D. R. Klein and H. Strandgaard. bibliog il J Wildlife Mgt 36:64-79

FIGURE 47. Biological and Agricultural Index. 1972 - 1973.

RIBONUCLEIC ACID, CHLORAMPHENICOL, NUCLEIC ACIDS
74/5-1755 PRECURSORS, RADIOACTIVITY, BIOSYNTHETIC STUDIES, CARBON MAGNETIC RESONANCE (CMR), CARBON-13, CARBON-14, DOUBLE LABELING
74/5-0373 PREDATION, SENSORY SIGNALS, SHARKS, TERRITORIALITY, ATTACK
74/5-0438 PREDATION, SAWFLY, INSECT, NATURAL DEFENSES, PEST CONTROL, PIN, PINE RESIN
74/5-0478 PREDATION, SHELL ACQUISITION, DOMINANCE HIERARCHY, GASTROPOD, HERMIT CRAB
74/5-1035 PREDATION, PREY, PROTOZOANS, COMMUNITY STRUCTURE, DIVERSITY, MOSQUITO (WYCOMIYA), PITCHER PLANT (SARRACENIA PURPUREA)
74/5-1576 PREDATION, RANGE, WILDLIFE, ENDANGERED SPECIES, MOUNTAIN LION, POPULATION STUDIES
74/5-1584 PREDATION, PREDATORY FISH, REEF, ANIMAL BEHAVIOR, CAMOUFLAGE, CAPTURE TECHNIQUES, FISH, MIMICRY
74/5-2625 PREDATION, RADIO TELEMETRY, SUPERIOR NATIONAL FOREST, WOLF, HOWLING, POPULATION BIOLOGY
74/5-2701 PREDATION, TRICHOMES, WEB-SPINNING, ADAPTIVE MECHANISMS, BUTTERFLY LARVAE, COEVOLUTION, DEFENSE MECHANISMS, MECHAMITIS ISTHMIA
74/5-3066 PREDATION, PREHISTORIC MAN, SOUTH AFRICA, CARNIVORES, EARLY MAN, EXTINCTION, HOMINID SKULLS, HOMINIDS, HUMAN EVOLUTION, LEOPARDS
74/5-3088 PREDATION, SOCIAL STRUCTURE, ANT COLONY, ANTS, BEHAVIORAL PATTERNS, HARVESTER ANTS, INSECT BEHAVIOR, INSECT COMMUNICATION, POGONOMYRMEX BADIUS
74/5-3140 PREDATION, PREDATOR-PREY SYSTEMS, PROTOZOA, COEXISTENCE, DIDINIUM NASUTUM, PARAMECIUM AURELIA, POPULATION DYNAMICS
74/5-3696 PREDATION, PRIMATE PREDATION, BABOONS, HUNTING STRATEGIES, OLIVE BABOONS
74/5-4704 PREDATION, RANGIFER TARANDUS, REPRODUCTIVE RATE, CARIBOU, CERVIDS, FOOD HABITS, HABITAT DESTRUCTION, HUNTING, MORTALITY RATE, POPULATION DECLINE
74/5-4658 PREDATION DETERRENTS, ASCLEPIAS CURASSAVICA, BUTTERFLIES, CARDIAC GLYCOSIDES, DANAIDAE, DANAUS PLEXIPPUS, DEFENSE MECHANISMS, EMETICS, MILKWEEDS, MONARCH BUTTERFLY
74/5-1041 PREDATOR, RANGE MANAGEMENT, CONSERVATION, COYOTE, ECOSYSTEM, FAMILY UNIT, POISONING
74/5-1042 PREDATOR, RED WOLF, WILDLIFE MANAGEMENT, WOLF, CANIS RUFUS, CONSERVATION, COYOTE, ENDANGERED SPECIES
74/5-2628 PREDATOR, COYOTE, EASTERN HABITAT, INTERBREEDING
74/5-2630 PREDATOR, SHEEP, WILD DOG, AUSTRALIA, DINGO, HERBIVORES
74/5-3686 PREDATOR, PREDATOR-PREY RELATIONSHIP, WILDLIFE MANAGEMENT, COYOTE, ECOLOGICAL BALANCE, POISON CAMPAIGNS
74/5-3119 PREDATOR CONTROL, WILDLIFE MANAGEMENT, BRITISH COLUMBIA, CANADA, POISON CAMPAIGN, POISONS
74/5-4707 PREDATOR CONTROL, STEEL TRAPS, STRYCHNINE, TARGET SPECIES, TRAPS, WILD TURKEY, WILDLIFE MANAGEMENT TECHNIQUES, BAIT TECHNIQUES, BOBCATS, CANIS LATRANS, COYOTES LYNX RUFUS, M-44'S
74/5-4706 PREDATOR REMOVAL, WHITE-TAILED DEER, WILDLIFE MANAGEMENT TECHNIQUES, BOBCAT PREDATION, CANIS LATRANS, COYOTE PREDATION, LYNX RUFUS, ODOCOILEUS VIRGINIANUS
74/5-4698 PREDATOR RESTORATION, TROUT, WILDLIFE MANAGEMENT, WOLVES, ANIMAL REINTRODUCTION, ELK, GRIZZLY BEARS, LAND MANAGEMENT, NATIONAL PARKS
74/5-3162 PREDATOR SPECIES, SOLITARY SPECIES, TERRITORIAL ANIMALS, WOLF,

TERRITORIAL ANIMAL, WILDLIFE MANAGEMENT, DEER, FOREST SHELTER
74/5-4174 PREDATOR-PREY RELATIONSHIP, REEF FISHES, TROPICAL FISH COLORATION, ADAPTIVE CHARACTERISTICS, CAMOUFLAGE, COMMUNICATION MECHANISMS, FISH COLORATION
74/5-0765 PREDATOR-PREY RELATIONSHIPS, PROTOZOA, REPRODUCTION, SIZE-SELECTIVE PREDATION, ZOOPLANKTON, DAPHNIA
74/5-1034 PREDATOR-PREY RELATIONSHIPS, PROTOZOA, TROPIC RELATIONSHIPS, DIDINUM NASUTUM, NATURAL CONTROLS, PARAMECIUM AURELIA
74/5-1563 PREDATOR-PREY RELATIONSHIPS, PREY LOCATION, BATS, ECHOLOCATION, HEARING, HORSESHOE BATS, INSECT CAPTURE
74/5-2627 PREDATOR-PREY RELATIONSHIPS, WILDLIFE MANAGEMENT, WILDLIFE PRESERVATION, WOLF, ISLE ROYALE, LAKE SUPERIOR, MOOSE
74/5-4201 PREDATOR-PREY RELATIONSHIPS, SENSORY DEVELOPMENT, VISION ADAPTATIONS, VISUAL CAPACITY, ADAPTIVE CHARACTERISTICS, NOCTURNAL ANIMALS, OWL
74/5-3140 PREDATOR-PREY SYSTEMS, PROTOZOA, COEXISTENCE, DIDINIUM NASUTUM, PARAMECIUM AURELIA, POPULATION DYNAMICS, PREDATION
74/5-0434 PREDATORS, PREY, SHEEP, CONDITIONING, COYOTES, LITHIUM CHLORIDE, POISON
74/5-0459 PREDATORS, COOPER'S HAWK (ACCIPITER COOPERII), DDT, HAWKS, PESTICIDES

FIGURE 48. Biology Digest. vol. 1, 1974 - 1975. Cumulative Index.

However, unlike *Biological Abstracts* this is not a key word in context index. Each article is assigned a number of subject terms taken from the abstract and the title which are listed in the index. Each entry includes not only the particular term under which the article is being listed, but also the other terms assigned to that article. Each entry serves, therefore, as a mini--summary of the contents of the article.

For your topic, FIGURE 29 indicates that the likely terms to search are predator (prey), deer (Cervidae) and all the deer types and their predators. A search in *Biology Digest* indicates that "Predation," "Predator," and "Predator--Prey" are possible terms to search. Once you have established the best terms, read each entry and eliminate those which because of the addi--tional terms that are listed appear not to be rele--vant to your topic. For example, in FIGURE 48, the entry that begins: "74/5--3696 Predation, Pri--mate Predation, Baboons," is clearly irrelevant because of the mention of "Baboons." On the other hand, the very next item looks useful be--cause of the terms "Rangifer tarandus." (Note here that the term "Caribou" is also listed.)

In FIGURE 48 the other useful articles are also marked. The entry that begins "74/5--4707 Predator Control, Steel traps" looks useful be--cause "Canis latrans" is mentioned. However, because "wild turkey" is also mentioned you can fairly safely conclude that the article will not be useful.

FIGURE 49 shows the abstract for one of the articles. It is illustrative of the particularly lengthy abstracts that *B.D.* includes.

Science Citation Index.
"Permuterm Index"

In chapter 5 we explained the use of the *Science Citation Index.* However we did not mention that a subject term approach is also possible through its "Permuterm Index." Because the "Permuterm Index" was sold as a supplement to *SCI* at a substantial additional cost there are a number of libraries which did not subscribe to it until 1978. If your library does not have *Bio--logical Abstracts* but does have *SCI*'s "Permuterm Index" you will want to consult it.

The "Permuterm Index" of *SCI* is a key word index that provides indexing by every two word combination derivable from the title. This is *not* limited to every two consecutive words, but every pair regardless of the relationship within the title.

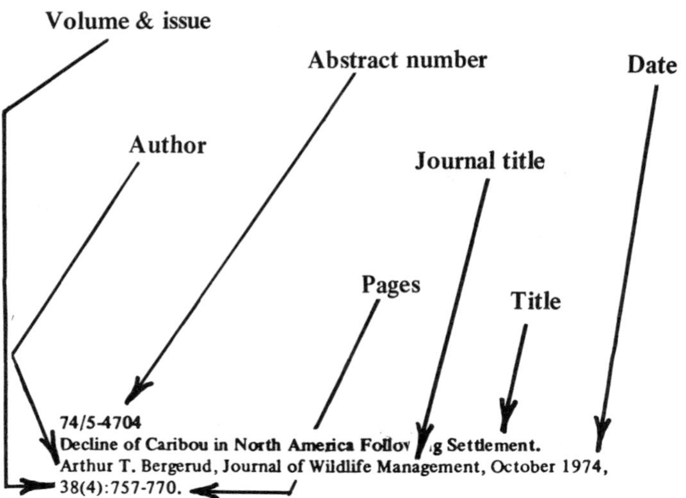

74/5-4704
Decline of Caribou in North America Following Settlement.
Arthur T. Bergerud, Journal of Wildlife Management, October 1974, 38(4):757-770.

The numbers of caribou (*Rangifer tarandus*) in North America generally declined in the late 1800s and early 1900s. Several hypotheses have been proposed to explain this population reduction. The first hypothesis is that caribou declined because of the destruction of lichen pastures by fire and logging; most northern biologists believe that man's destruction of caribou habitat was the main cause of the decline. Supporters of this hypothesis assume that caribou require lichens in the winter. Actually, however, caribou almost invariably lose weight on a diet of only lichens. Caribou do not require lichens in the winter and, in many areas, mainly eat other foods (e. g., balsam, fir, juniper, and sedges); caribou are apparently more versatile in their feeding habits than other North American cervids (deer, elk, moose, and related animals). Additionally, there is no substantial evidence that fires increased in northern Canada with early settlement. Another weakness of the habitat hypothesis is that no one has documented that a reduction in absolute abundance of forage has affected either the reproductive rate or the mortality rate of free-ranging caribou. There is little evidence to date that large numbers of free-ranging caribou have starved to death in winter in North America.

The decline of the caribou has also been attributed to overhunting. Caribou are more vulnerable to hunting than other North American cervids. They travel in large herds over vast areas, making them easily accessible to hunters. It has been hypothesized that at the time of the settlement of North America, there was a fine balance between gains and losses in caribou populations. The caribou's reproductive rate is relatively low, and they were hunted by a versatile predator–the wolf. With the advent of hunting with rifles, the precarious balance between population recruitment and mortality was upset, and populations began to decline. The reproductive rate of caribou is low, but the mortality rate among calves is high; many factors are involved in this mortality including birth defects, accidents, disease, social interactions, chilling of the young, and predation. It was hypothesized in this review that predation (natural and by man) is the most serious and consistent of these mortality factors. It was also suggested that natural predation may have increased in the early 1900s in areas such as British Columbia and Ontario where habitats were altered by fire and logging. With settlement of regions inhabited by the caribou, there was increased hunting and, in some areas, more natural predation losses (possibly also increased disease). This increased mortality was sufficient to initiate a decline between reproduction and natural mortality in undisturbed populations.

The third hypothesis is a combination of the first two. The fourth hypothesis is that caribou declined in Alaska due to increased movement to marginal habitats when the numbers were high because increased densities provided a "social stimulus" leading to population shifts. This hypothesis, proposed to explain declines in caribou herds in Alaska in the 1880s and 1940s has not been substantiated by evidence of large-scale die-offs in these periods. This review supported the second hypothesis–that caribou populations declined because of increased hunting mortality and natural predation of some herds.

FIGURE 49. Biology Digest. vol. 1, May, 1975.

FIGURE 50 shows an example of a search under "Deer" as the first word, and the "modi--fying" words "Population," and "Populations." After each word combination is the author's name. This must be checked in the "Source Index" portion of *SCI* (FIGURE 51), for the bibliographic information needed to ascertain if the article is useful and to locate it. This ex--ample illustrates beautifully the problem of key word indexes like *Biological Abstracts,* and par--ticularly the "Permuterm Index." Of the seven articles under "Deer Population(s)," only two of those appear useful. Several are on deer mice, and the others are on aspects of deer population irrele--vant to our topic.

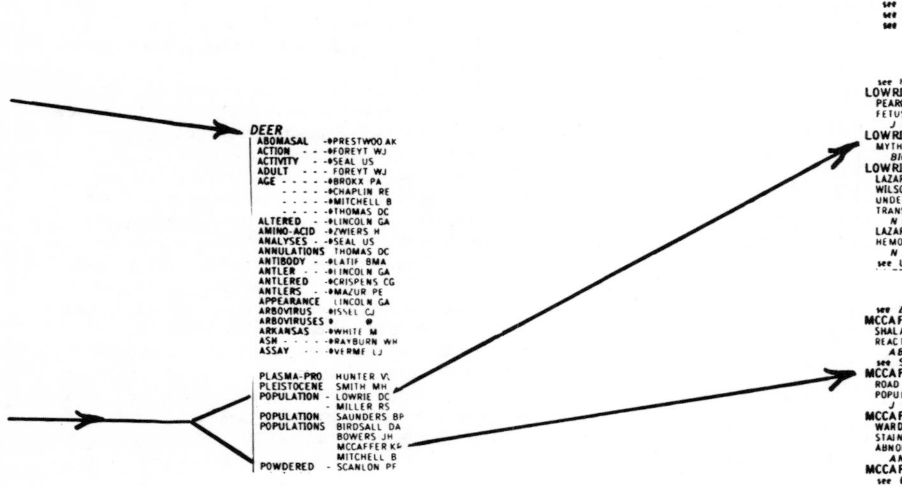
DEER
ABOMASAL -•PRESTWOO AK
ACTION - - -•FOREYT WJ
ACTIVITY - -•SEAL US
ADULT - - - FOREYT WJ
AGE - - - - - -•BROKX PA
- - - - - -•CHAPLIN RE
- - - - - -•MITCHELL B
- - - - - -•THOMAS DC
ALTERED - -•LINCOLN GA
AMINO-ACID -•ZWIERS H
ANALYSES - -•SEAL US
ANNULATIONS THOMAS DC
ANTIBODY - -•LATIF BMA
ANTLER - - -•LINCOLN GA
ANTLERED -•CRISPENS CG
ANTLERS - -•MAZUR PE
APPEARANCE LINCOLN GA
ARBOVIRUS •ISSEL CJ
ARBOVIRUSES • •
ARKANSAS -•WHITE M
ASH - - - - - -•RAYBURN WH
ASSAY - - -•VERME LJ

PLASMA-PRO HUNTER VL
PLEISTOCENE SMITH MH
POPULATION - LOWRIE DC
- MILLER RS
POPULATION SAUNDERS BP
POPULATIONS BIRDSALL DA
BOWERS JH
MCCAFFER KR
MITCHELL B
POWDERED - SCANLON PF

FIGURE 50. Science Citation Index. "Permuterm Index," 1973.

"But which index do I use?"

As a biology major doing scholarly library research you are primarily interested in getting the most current research literature. Therefore, *Biological Abstracts* is the preferred index. However, your library may not have it, and you will want to use *Biological & Agricultural Index, Biology Digest,* or the "Permuterm Index" of *Science Citation Index.* Which you use depends upon what your topic is. In this case, predator--prey relationships, any of the three is appropriate. However, if the subject is physiological/biochemical in nature, or is in the area of agriculture, the *B.& A.I.* or *SCI* are more useful. In still other areas such as medi--cine, environmental sciences and specific subdis--ciplines of the biological sciences, other indexes would be more important supplements or alterna--tives to searching *Biological Abstracts*. It is not our

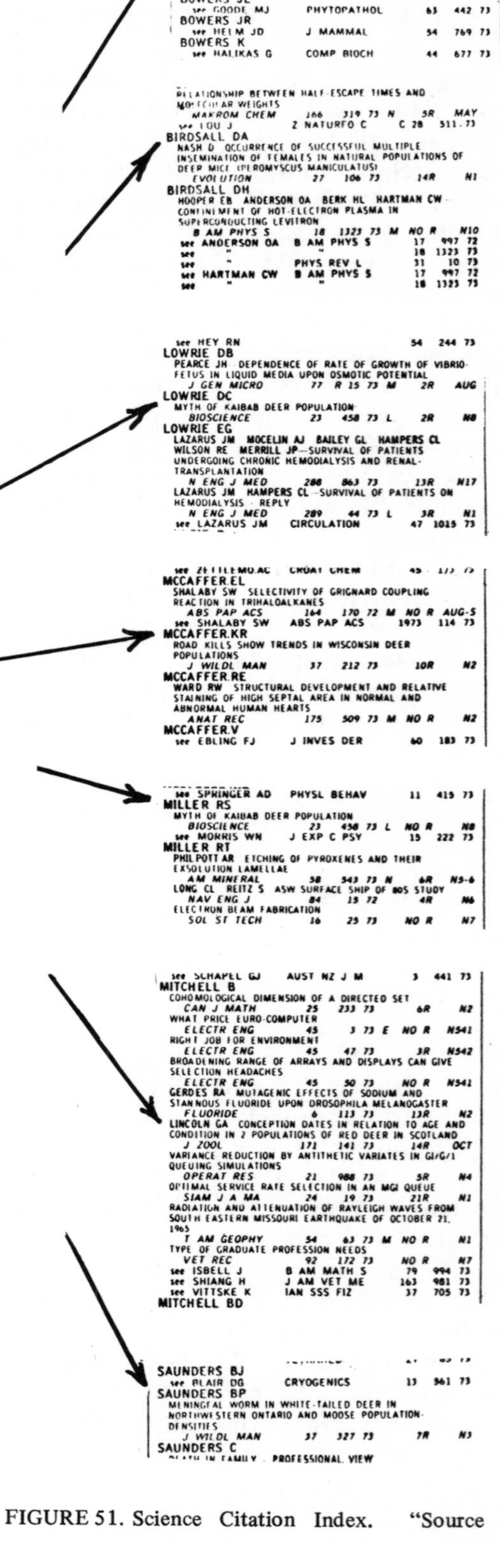
see PSYCHOYO.S ENDOCRINOL 92 243 73
BOWERS J
see DOMINO EF ARCH G PSYC 29 195 73
see " BRAIN RES 58 179 73
BOWERS JA
see JOHNSON PG POULTRY SCI 52 2046 73
see PENNER KK J FOOD SCI 38 583 73
BOWERS JH
BAKER RJ SMITH MH CHROMOSOMAL, ELECTROPHORETIC, AND BREEDING STUDIES OF SELECTED POPULATIONS OF DEER MICE (PEROMYSCUS MANICULATUS) AND BLACK EARED MICE (P MELANOTIS)
EVOLUTION 27 378 73 32R N3
BOWERS JL
see GOODE MJ PHYTOPATHOL 63 442 73
BOWERS JR
see HELM JD J MAMMAL 54 769 73
BOWERS K
see HALIKAS G COMP BIOCH 44 677 73

RELATIONSHIP BETWEEN HALF-ESCAPE TIMES AND MOLECULAR WEIGHTS
MAKROM CHEM 166 319 73 N 5R MAY
see LOU J Z NATURFO C C 28 511 73
BIRDSALL DA
NASH D OCCURRENCE OF SUCCESSFUL MULTIPLE INSEMINATION OF FEMALES IN NATURAL POPULATIONS OF DEER MICE (PEROMYSCUS MANICULATUS)
EVOLUTION 27 106 73 14R N1
BIRDSALL DH
HOOPER EB ANDERSON OA BERK HL HARTMAN CW CONFINEMENT OF HOT-ELECTRON PLASMA IN SUPERCONDUCTING LEVITRON
B AM PHYS S 18 1323 73 M NO R N10
see ANDERSON OA B AM PHYS S 17 997 72
see " " 18 1323 73
see " PHYS REV L 31 10 73
see HARTMAN CW B AM PHYS S 17 997 72
see " " 18 1323 73

see HEY RN 54 244 73
LOWRIE DB
PEARCE JH DEPENDENCE OF RATE OF GROWTH OF VIBRIO-FETUS IN LIQUID MEDIA UPON OSMOTIC POTENTIAL
J GEN MICRO 77 R 15 73 M 2R AUG
LOWRIE DC
MYTH OF KAIBAB DEER POPULATION
BIOSCIENCE 23 458 73 L 2R N8
LOWRIE EG
LAZARUS JM MOCELIN AJ BAILEY GL HAMPERS CL WILSON RE MERRILL JP—SURVIVAL OF PATIENTS UNDERGOING CHRONIC HEMODIALYSIS AND RENAL-TRANSPLANTATION
N ENG J MED 288 863 73 13R N17
LAZARUS JM HAMPERS CL—SURVIVAL OF PATIENTS ON HEMODIALYSIS - REPLY
N ENG J MED 289 44 73 L 3R N1
see LAZARUS JM CIRCULATION 47 1015 73

MCCAFFER.EL
SHALABY SW SELECTIVITY OF GRIGNARD COUPLING REACTION IN TRIHALOALKANES
ABS PAP ACS 164 170 72 M NO R AUG-5
see SHALABY SW ABS PAP ACS 1973 114 73
MCCAFFER.KR
ROAD KILLS SHOW TRENDS IN WISCONSIN DEER POPULATIONS
J WILDL MAN 37 212 73 10R N2
MCCAFFER.RE
WARD RW STRUCTURAL DEVELOPMENT AND RELATIVE STAINING OF HIGH SEPTAL AREA IN NORMAL AND ABNORMAL HUMAN HEARTS
ANAT REC 175 509 73 M NO R N2
MCCAFFER.V
see EBLING FJ J INVES DER 60 183 73

see SPRINGER AD PHYSL BEHAV 11 415 73
MILLER RS
MYTH OF KAIBAB DEER POPULATION
BIOSCIENCE 23 458 73 L NO R N8
see MORRIS WN J EXP C PSY 15 222 73
MILLER RT
PHILPOTT AR ETCHING OF PYROXENES AND THEIR EXSOLUTION LAMELLAE
AM MINERAL 58 543 73 N 6R N5-6
LONG CL REITZ S ASW SURFACE SHIP OF 80S STUDY
NAV ENG J 84 15 72 4R N6
ELECTRON BEAM FABRICATION
SOL ST TECH 16 25 73 NO R N7

see SCHAPEL GJ AUST NZ J M 3 441 73
MITCHELL B
COHOMOLOGICAL DIMENSION OF A DIRECTED SET
CAN J MATH 25 233 73 6R N2
WHAT PRICE EURO-COMPUTER
ELECTR ENG 45 3 73 E NO R N541
RIGHT JOB FOR ENVIRONMENT
ELECTR ENG 45 47 73 3R N542
BROADENING RANGE OF ARRAYS AND DISPLAYS CAN GIVE SELECTION HEADACHES
ELECTR ENG 45 50 73 NO R N541
GERDES RA MUTAGENIC EFFECTS OF SODIUM AND STANNOUS FLUORIDE UPON DROSOPHILA MELANOGASTER
FLUORIDE 6 113 73 13R N2
LINCOLN GA CONCEPTION DATES IN RELATION TO AGE AND CONDITION IN 2 POPULATIONS OF RED DEER IN SCOTLAND
J ZOOL 171 141 73 14R OCT
VARIANCE REDUCTION BY ANTITHETIC VARIATES IN GI/G/1 QUEUING SIMULATIONS
OPERAT RES 21 988 73 5R N4
OPTIMAL SERVICE RATE SELECTION IN AN MGI QUEUE
SIAM J A MA 24 19 73 21R N1
RADIATION AND ATTENUATION OF RAYLEIGH WAVES FROM SOUTH EASTERN MISSOURI EARTHQUAKE OF OCTOBER 21, 1965
T AM GEOPHY 54 63 73 M NO R N1
TYPE OF GRADUATE PROFESSION NEEDS
VET REC 92 172 73 NO R N7
see ISBELL J B AM MATH S 79 994 73
see SHIANG H J AM VET ME 163 981 73
see VITTSKE K IAN SSS FIZ 37 705 73
MITCHELL BD

SAUNDERS BJ
see BLAIR DG CRYOGENICS 13 361 73
SAUNDERS BP
MENINGEAL WORM IN WHITE-TAILED DEER IN NORTHWESTERN ONTARIO AND MOOSE POPULATION-DENSITIES
J WILDL MAN 37 327 73 7R N3
SAUNDERS C
DEATH IN FAMILY - PROFESSIONAL VIEW

FIGURE 51. Science Citation Index. "Source Index," 1973.

purpose to describe and explain all of the dozens of indexes which a biologist might use. The most important specialized indexes are included in Appendix IV, a bibliography of reference sources useful for library work in specific biology courses.

Two final comments about the selection of the appropriate index are in order. In the area of biochemistry, or whenever a topic has a chemical dimension *Chemical Abstracts* should be used. For this sample topic *Chemical Abstracts* was not appropriate. But for many others not using *C.A. would be a serious oversight;* it is *the* most important index for any topic of a chemical nature. For this reason an appendix specifically on *C.A.* has been included in this guide, and Appendix V should be read and you should consult with a reference librarian before using *C.A.*

Second, there is an index to the literature on specific animal groups. This index, *Zoological Record,* is an important one for searches like the one on which you are working here. However, it is not held by most small college libraries and it partially duplicates *Biological Abstracts.* Therefore, *Zoological Record* has been treated separately in Appendix VI.

* * *

Summary

1. There are several indexes to the biological journals, of which *Biological Abstracts* is the most important.
2. *Biological Abstracts* consists of bibliographical citations plus summaries for each article. These summaries are accessible through one of five indexes: Author, Subject, Generic, Biosystematic, and CROSS (general topic).
3. The Subject, Biosystematic, and CROSS indexes are complex and therefore require a substantial understanding of your topic and knowledge of the organization of the indexes.
4. *Bioresearch Index,* companion to *B.A.*, covers additional literature that is not in *B.A.*
5. When *B.A.* is not available *Biological and Agricultural Index* and/or *Biology Digest* should be used as an index to the biological literature. The "Permuterm Index" of the *Science Citation Index* is another useful index if *B.A.* is not available.
6. For topics of a chemical nature, *Chemical Abstracts* is the most useful. See Appendix V.
7. For more complete indexing of the literature on specific animal groups *Zoological Record* should be used. It is explained in Appendix VI.
8. There are many specialized indexes to subdisciplines of biology. These are included in Appendix IV.

"Why should I be concerned about the most recent publications?"

Science is ever changing. While many ideas have been around for a long time, there are many concepts and much experimental evidence to support those concepts which are of recent origin. In some fields of biology developments are occurring so rapidly that our understanding of particular phenomena may change in just a few months with the publication of significant new research findings. It is therefore necessary to get very recent material in order to understand the currently accepted theories, or at least to understand the contradictory lines of evidence.

A less pompous reason, but of significance to you, is the higher regard with which a professor will hold a paper which cites very recent papers. (You might even educate the professor by using sources of which s/he is not aware!)

But isn't use of the latest issue of *Biological Abstracts* or other indexes mentioned in Chapters 5 and 6 enough? Generally the answer to that question is no. The time between publication of an issue of a journal and the publication of an index which covers it can be anywhere from eight weeks to a year or more. While indexes like *Biological Abstracts* have made monumental efforts in recent years to reduce the gap, there remains a time period, generally about six months, between the literature covered in the most recent issue of an index and the time at which you are doing the search.

"If there is no index, how can I find this new material?"

The most obvious way is to simply browse through the last few issues of those journals you have already found most useful. This has two limitations, however. One, it is inconvenient to browse through dozens of issues of several different titles. Two, you may not have immediate access to all the important journals you want.

The solution to these problems is *Current Contents,* which serves an indexing function, but is organized and specifically designed to give rapid access to current literature, and not to serve as a permanent retrospective indexing tool. That is why *CC* is referred to as a "current awareness service."

Current Contents is published by the Institute for Scientific Information (the same company that publishes *Science Citation Index*). There actually are six different subject clusterings each covering between 700 and 1100 journals, and with some overlap amongst them. The subjects are:

Agriculture, Biology and Environmental Sciences
Social and Behavior Sciences
Clinical Practice
Engineering, Technology and Applied Sciences
Life Sciences
Physical and Chemical Sciences

Each of these is published weekly and includes copies of the Tables of Contents of journals in the general field (FIGURE 52). These Tables of Contents generally are six weeks or less old when they appear in *CC.* Each issue of *CC* contains an "Author Index" (FIGURE 53) and, depending on which edition your library gets, a "Subject Index" (FIGURE 54). Each issue of *CC* also lists all journals and their issue numbers for which contents are included in that issue.

Using Current Contents

The first problem you face is which *CC* to use. This can be quickly solved by looking at the list of journals included, published periodically in each series, to see which series covers the journals you have found useful. In the case of our topic on the role of predators in the control of North American deer population size, both or either "Agriculture, Biology, and Environmental Sciences," and "Life Sciences" could be used.

Once the appropriate section has been chosen, there are three possible approaches. In each case you should begin with the most recent issue of *CC* and go backwards until you begin relocating older articles of which you are already aware. The three

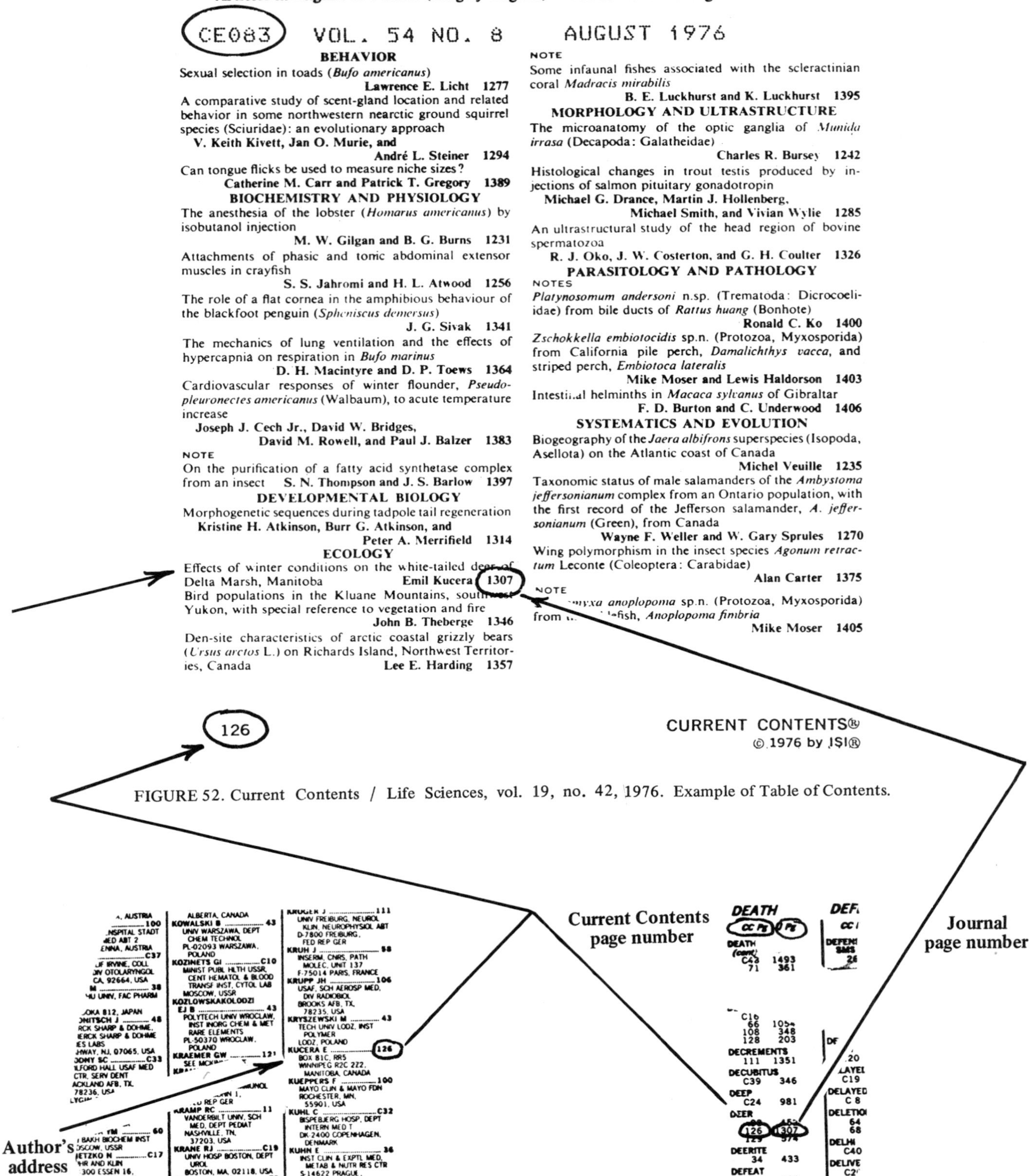

Canadian Journal of Zoology

Articles in English or French (Largely English)–Each Abstract in English and French

CE083 VOL. 54 NO. 8 AUGUST 1976

126

CURRENT CONTENTS®
© 1976 by ISI®

FIGURE 52. Current Contents / Life Sciences, vol. 19, no. 42, 1976. Example of Table of Contents.

FIGURE 53. Current Contents / Life Sciences, vol. 19, no. 42, 1976. Author Index.

FIGURE 54. Current Contents / Life Sciences, vol. 19, no. 42, 1976. Subject Index.

approaches are:

1) Search the "Author Index" (FIGURE 53) for names of workers in your area.
2) Use the list of journals covered in each issue to identify where Tables of Contents of the important journals for your topic are located and then browse those contents.
3) If your library has the edition which includes the "Subject Index" (FIGURE 54) you can search that index for key terms.

Summary

1. Rapid developments in science make it important to find the latest information.
2. Index coverage lags between eight weeks and a year behind the publication of the journals.
3. *Current Contents*' six different series provide access to recent issues of journals through the republication of their Tables of Contents and accompanying weekly journal title, author, and subject indexes, and occasional listings of journals covered.
4. *Current Contents* can be searched using the names of important authors in a field, or by browsing issues that contain Tables of Contents for key journals in the field.
5. If the library has the edition of *CC* which has the key word subject index you can search by subject.

Why Use a Guide?

This book is a *selective* guide to the basic sources in biology, and therefore it may not answer all your questions. Especially if you are an advanced student, you may need a more comprehensive guide in order to find specialized reference sources which are uniquely useful for your topic. Guides to the literature of a subject and comprehensive lists of reference sources can lead to materials in your library and can also point to the existence of materials not in your library collection. Reference librarians rely heavily on these guides and lists, and if finding enough material is a problem, you should be aware of such guides also. The main problem with comprehensive guides is selecting the few important sources from the many that are cited.

How to Use a Guide in Biology

The most useful comprehensive guide to reference sources in biology is R.T. Bottle's *The Use of Biological Literature* (2nd ed., Handen, Conn., Shoe String Press, 1971). In addition to chapters on the various types of literature sources -- journals, serials, patents, indexes, bibliographies, and handbooks -- there are individual sections on major subject areas of interest to biologists: botany, zoology, ecology, genetics, biochemistry and biophysics, microbiology, medicine, and food and agriculture. In each case there are extensive lists of specific titles. One of Bottle's helpful features is the "Quick Index of Subjects" and a "Quick Index for Indexing and Abstracting Services," which appears immediately after the Table of Contents. FIGURE 55 shows the "Quick Index of Subjects" which illustrates that our topic, animal ecology, is discussed on page 197. There we are given some helpful information on the literature of mammalian ecology. However it is also important to check the earlier portions of the "Ecology" chapter for texts, indexes and other reference sources. For example, FIGURE 56 shows the section discussing reviews. Such titles as *Ecological Studies* and *Advances in Trop--*

QUICK INDEX OF SUBJECTS

The main entry for each subject is given, other entries and subjects will be found in the full index.

Agriculture, 298
Animals, laboratory, 281

Bacteriology, 268

Ecology,
general, 188
animal, 197
plant, 202
Embryology, 180
Endocrinology, 297
[illegible]
Food,
general, 298
legal, 306
microbiology, 279

Genetics, 206

Histology, 187
History, 311

Immunology, 275
Invertebrates, 172

Laboratories,
animals, 281
design, 336
safety, 336
Law, food, 306

Patents, 50
Pathology, 289
Pharmacology, 292
Pharmacy, 121
Physiology,
animal, 181
human, 295
[illegible]
Teratology, 292
Tissue Culture, 274
Toxicology, 292
Translations, 37

Vertebrates, 177
Virology, 271

Zoology, 169

FIGURE 55. Bottle, R. T. The Use of Biological Literature.

Reviews

Review serials, like Symposia, are usually l[...]d to particular aspects of ecology. The exception is th[e a]nnual *Advances in Ecological Research* edited by J. B. C[...], each volume of which contains three or four reviews in [de]pth on limited but widely different topics. Ecological Studies (Springer Verlag) is a new series of volumes, each one reviewing a subject in depth. *Ecological Reviews Tohuku* (Japan) should also be mentioned, as well as *Recent Advances in Tropical Ecology* edi[ted] by R. Misra and B. Gopal (International Society for Tropical Ec[...]. [...]gical reviews appear from time to time in a number of journals ... *Quarterly Reviews of Biology, Biological Reviews, Botanical Reviews, Advances in Botanical Research, Recent Advances in Botany, Viewpoints in Biology* and *Symposia on Quantitative Biology*. *Monographiae Biologicae* (W. Junk) publish collections of papers on particular problems, e.g. crop pests or concerning a particular region, e.g. South America, Lake Ohrid: these are largely ecological in content and comprehensive in coverage. *Travaux du*

FIGURE 56. Bottle, R. T. The Use of Biological Literature, p. 190.

ical Ecology were not discussed earlier in this guide.

Two other guides to the literature of biology should be mentioned. R.C. Smith, *Guide to the Literature of the Life Sciences,* 8th ed. (Minneapolis, Burgess Publishing Corp., 1972) is more selective than Bottle, except for keys, catalogues, and checklists of specific plant and animal groups where Smith has an excellent listing. A.E.Kerker's *Biological and Biomedical Resource Literature* (LaFayette, Ind., Purdue University, 1968), while substantially out of date, is still a good listing of reference sources in the fields of biology and research medicine other than surgery and medical technology.

How to Use General Reference Sources

The most comprehensive annotated bibliographies of reference sources in English are: A.J. Walford, *Guide to Reference Material,* 3rd ed., (London, The Library Association, 1973--), and Eugene P. Sheehy, *Guide to Reference Books,* 9th ed., (Chicago, American Library Association, 1976). Volume 1 of Walford has a fifty--eight page section on general biology, botany and zoology, and other sections on related aspects. Walford and Sheehy can be useful as is demonstrated in the sample entry from Sheehy shown in FIGURE 57. Here

the literature for 1895. The individual volumes have no indexes, but the *Bibliographia zoologica* is indexed in the *Register* of the *Zoologischer Anzeiger,* Jahrg. 16–40, 1899–1922 (5v.). Z7991.B6

Wildlife review. no.67– , June 1952– . [Laurel, Md., 1952–] Quarterly. EC136

Frequency varies.

Issued currently by U.S. Sport Fisheries and Wildlife Bureau; formerly issued by U.S. Interior Dept. Fish and Wildlife Service.

A bibliography of the literature of wildlife management, arranged by subject. 25 to 50 percent of the entries include abstracts or references to abstracts. Each issue has an author index. SK351.W58

FIGURE 57. Sheehy, Eugene P. Guide to Reference Books, p. 729.

we find that there is a specialized index, *Wildlife Review*. According to the *Guide*'s editors, *W.R.* might be a good supplement to *Biological Abstracts* or the other biology indexes.

* * *

Summary

1. Guides to the literature of biology may help you locate specialized materials useful for your paper.
2. Guides to the literature of science, biology and chemistry, appear in Appendix III.

9 Using Other Libraries

It is satisfying to use bibliographies, periodical indexes, and abstracts when they lead you to vital books and articles in your library. However, these reference sources can also be frustrating if, after locating some choice items, you find that your library does not own them. This happens to both students and faculty even at the largest libraries. Fortunately the problem is not insoluble *if you act in time.*

How to Request Materials from Other Libraries

Your reference librarian may be able to borrow the books you need from another library, and s/he may be able to get you photocopies of any articles you need. All s/he needs is time, a full and accurate citation and, in the case of photocopying, some cash. The time required varies from a few days to a few months, depending on whether your library is part of a cooperating network of libraries and whether the material is available in the library where it is requested. An average wait is about two weeks. The full citation generally includes a report of the page where you found the book or article cited. This may seem like bureaucratic red tape, but this information is required by lending libraries and is good insurance against errors in transmission. If you do not have the full bibliographical information, ask your reference librarian to help you find it. The cost of photocopies is generally ten cents or more per exposure. Libraries often make no charge for mailing books, but they generally do not mail periodicals.

If you are an undergraduate at a university, you may find that this interlibrary loan service is not available to you, partly because the interlibrary loan code forbids requests for undergraduates and partly because a university library serving doctoral students is presumed to have a collection that is adequate for undergraduates. Interlibrary loan service is more readily available to undergraduates at colleges which have made special arrangements to borrow from a nearby university or state library.

Visiting Another Library

If a large university is close, your time is short, or your library will not borrow for you, then you may prefer to visit another library. Your reference librarian can give the address, phone number, subject specialties and perhaps the hours of most libraries you may want to visit. You might ask which are closest and whether you will need a letter of introduction or a pass in order to use them. Most libraries let visitors use materials in the library only.

* * *

Summary

1. Bibliographies and periodical indexes can be frustrating to use if your library does not own the materials cited.
2. You can ask your reference librarian to borrow books from other libraries or to get photocopies of articles. Allow enough time for this procedure.
3. You can also visit other libraries, with help from your reference librarian.

APPENDIX I

LIBRARY USE QUIZ

A. Directions: Use this catalog card to answer the questions below.

1. Would this card be filed under "P," "E," "Q," or "T?"
2. What is the title of the series?
3. How many pages are in the book?
4. Under which of the following would other cards for this book be found in the catalog?
   a. Ehrlich, Paul. yes no
   b. Holm, Richard W., joint author. yes no
   c. Evolution yes no
   d. McGraw–Hill series in population biology. yes no
   e. McGraw--Hill yes no

The process of evolution.

Sci
QH
366.2
E3.5
1974

Ehrlich, Paul R

The process of evolution [by] Paul R. Ehrlich, Richard W. Holm [and] Dennis R. Parnell. 2d ed. New York, McGraw-Hill [1974]

xv, 378 p. illus. 24 cm. (McGraw-Hill series in population biology)

Includes bibliographies.

1. Evolution. 2. Genetics. I. Holm, Richard W., joint author. II. Parnell, Dennis R., joint author. III. Title.
[DNLM: 1. Evolution. 2. Genetics. QH 366 E33p 1975]

QH366.2.E35 1974 575 74-3030
ISBN 0-07-019133-6 MARC

Library of Congress 74 [4]

FIGURE A. Catalog card.

* * *

B. Directions: Use this excerpt from the *Readers' Guide* to answer the questions below.

5. How would you determine the title of the periodical which has the article, "Diabetes drugs and fatal heart disease?"
6. What is the date of this article?
7. On what page(s) is it?
8. In what volume is it?
9. Under what subject heading will you find similar articles?
10. What are the terms "antibiotics," "anti--depressants," etc.?"

* * *

Solubilization of a stereospecific opiate-macromolecular complex from rat brain. E. J. Simon and others. bibl il Science 190: 389-90 O 24 '75

DRUG research. See Pharmaceutical research

DRUG stores. See Drugstores

DRUG testing. See Drugs—Testing

DRUGS

Ceremonial chemistry, by T. Szasz. Review
Psychol Today 8:88+ F '75. S. H. Snyder

How safe are your prescription drugs? D. S. Greenberg. il Sci Digest 77:50-5 Je '75

Pills, profits, and politics, by M. Silverman and P. R. Lee. Review
Psychol Today 8:96+ Ap '75. R. E. Ornstein

What the British can prescribe. Time 106: 53 S 29 '75

See also

Antibiotics
Antidepressants
Cancer inhibiting substances
Doping in sports
Fertility drugs
Narcotic antagonists
Placebos
Tranquilizing drugs
United States—Food and drug administration
also names of drugs, e.g. Cocaine

Advertising

See Drug industry—Advertising; Drugstores—Advertising

Metabolism

How food and drugs fight it out. J. Arehart-Treichel. Sci N 108:43 Jl 19 '75

Physiological effects

Childbirth drugs and your baby. J. E. Brody. il Redbook 145:82-3+ Ag '75

Danger ahead! Valium: M. Nyswander's report. M. Nyswander. Vogue 165:152-3 F '75

Designing the safer drug; work of G. Banker. Sci Digest 77:8 My '75

Detection of mutagenic activity of metronidazole and niridazole in body fluids of humans and mice. M. S. Legator and others. bibl il Science 188:1118-19 Je 13 '75

Diabetes drugs and fatal heart disease. Sci N 107:69-70 F 1 '75

Do you take Valium? Selling of Valium. D. Larned. Ms 4:26-8+ N '75

Growing debate over safety of drugs. il U.S. News 78:61-2 Je 16 '75

FIGURE B. Readers' Guide excerpt.

Answers to the Library Use Quiz

A.

1. "P" This is a replica of the "title card" that would be filed in the catalog under "Process of evolution." When the initial word is "a," "an," or "the," it is ignored.
2. McGraw-Hill series in population biology.
3. 378 pages plus 15 introductory pages. Since the introductory pages are numbered differently they are listed separately.
4. a. yes. He is the author of the book.
   b. yes. He is a joint author. When there are three or less authors of a book, there will be a card for the book under each author.
   c. yes. This is one of the two subject headings under which Ehrlich's book is listed.
   d. no. Although the book is part of a series, the series name is not used as an entry under which the book is listed. You can tell this because the numbered items at the bottom of the card do not include "IV. Series." Many series, particularly if the individual parts are numbered, will have a series entry.
   e. no. Rarely is the publisher given an entry. If it were, there would be a numbered item at the bottom of the card (e.g. V. McGraw-Hill).

B.

5. Look in the list of periodicals indexed at the front of the volume. Most indexes have such a list which decodes the abbreviated titles used in the index. In this case the title is *Science News.*
6. February 1, 1975.
7. Pages 69--70.
8. Volume 107.
9. "Antibiotics," or one of the other "see also" references which is appropriate to the specific topic on which you are working.
10. "Antibiotics," etc., are subject headings where related and more specific aspects of the general topic "Drugs" can be located.

APPENDIX II

List of Review Serials

Outline

1. Biology, General
1B. Botany, General
1Z. Zoology, General
2. Plant Physiology (See also Botany, General)
3. Behavior
4. Genetics
5. Ecology (See also Marine Biology)
6. Biochemistry and Biophysics (See also Cell and Molecular Biology, Plant Physiology, and Animal Physiology)
7. Cell and Molecular Biology (See also Bio-chemistry and Biophysics, Plant Physiology, and Animal Physiology)
8. Evolution
9. Animal Physiology; Comparative Physiology; Comparative Anatomy (See also Biochemistry and Biophysics, and Cell and Molecular Biology; for Human Physiology see also Human Physiology and Medicine)
10. Embryology; Development
11. Microbiology; Bacteriology; Virology (For Microscopy see also Biology, General)
12. Lower Plants; Algae; Fungi (Mycology); Mosses; Ferns
13. Higher Plants; Angiosperms; Gymnosperms
12. Invertebrates
13. Vertebrates; Herpatology; Icthyology; Mam-malogy; Ornithology
14. Parasitology
15. Marine Biology
16. Agriculture
17. Environmental Science; Pollution
18. Medicine

* * *

1. Biology, General

American Scientist, vol. 1–. New Haven, Conn.: Society of the Sigma XI, 1913--.

Biological Reviews, vol. 1–. Cambridge: University Press, 1923–.

Quarterly Review of Biology, vol. 1--. Stony Brook, New York: Stony Brook Foundation, Inc., 1926--.

Scientific American, vol. 1--. New York: Scientific American, 1845--.

1B. Botany, General

Advances in Botanical Research, vol. 1–. New York: Academic Press, 1963--.

Botanical Review, vol. 1–. New York: New York Botanical Garden, 1935–.

1Z. Zoology, General

International Review of General and Experimental Zoology, vol. 1–. New York: Academic Press, 1964--.

2. Plant Physiology (See also Botany, General)

Annual Review of Phytopathology, vol. 1--. Palo Alto, Cal.: Annual Reviews, 1963--.

Annual Review of Plant Physiology, vol. 1--. Palo Alto, Cal.: Annual Reviews, 1950--.

3. Behavior (Including Psychology)

Advances in Psychobiology, vol. 1--. New York: Wiley--Interscience, 1972--.

Advances in the Study of Behavior, vol. 1--. New York: Academic Press, 1965--.

Annual Review of Psychology, vol. 1--. Palo Alto, Cal.: Annual Reviews, 1951--.

4. Genetics

Advances in Genetics, vol. 1--. New York: Academic Press, 1947--.

Advances in Human Genetics, vol. 1--. New York: Plenum, 1970--.

Annual Review of Genetics, vol. 1--. Palo Alto, Cal.: Annual Reviews, 1967--.

5. Ecology (See also Marine Biology)

Advances in Ecological Research, vol. 1--. New York: Academic Press, 1962--.

Annual Review of Ecology and Systematics, vol. 1--. Palo Alto, Cal.: Annual Reviews, 1970--.

6. Biochemistry and Biophysics (See also Cell and Molecular Biology, Plant Physiology, and Animal Physiology)

Advances in Biological and Medical Physics, vol. 1–. New York: Academic Press, 1948--.

Advances in Biophysics, vol. 1–. Tokyo: University of Tokyo Press, 1970–.

Advances in Carbohydrate Chemistry and Biochemistry, vol. 1–. New York:

Academic Press, 1945–.

Advances in Enzymology and Related Areas of Molecular Biology, vol. 1–. New York: Interscience, Wiley, 1941–.

Advances in Lipid Research, vol. 1–. New York: Academic Press, 1963–.

Advances in Microbial Physiology, vol. 1–. New York: Academic Press, 1967–.

Advances in Protein Chemistry, vol. 1–. New York: Academic Press, 1944–.

Advances in Radiation Biology, vol. 1–. New York: Academic Press, 1964–.

Annual Review of Biochemistry, vol. 1–. Palo Alto, Cal.: Annual Reviews, 1932–.

Critical Reviews in Biochemistry, vol. 1–. Cleveland, Ohio: CRC Press, 1972–.

Essays in Biochemistry, vol. 1–. New York: Academic Press, 1965–.

Progress in Biophysics and Molecular Biology, vol. 1–. Oxford, Eng.: Pergamon Press, 1950–.

Progress in Nucleic Acid Research and Molecular Biology, vol. 1–. New York: Academic Press, 1963–.

Quarterly Reviews of Biophysics, vol. 1–. London: Cambridge University Press, 1968–.

Vitamins and Hormones, Advances in Research and Application, vol. 1–. New York: Academic Press, 1943–.

7. Cell and Molecular Biology (See also Biochemistry and Biophysics, Plant Physiology, and Animal Physiology)

Advances in Cell and Molecular Biology, vol. 1–. New York: Academic Press, 1971–.

Advances in Cell Biology, vol. 1–. Englewood Cliffs, N.J.: Prentice-Hall, 1970–.

Current Topics in Cellular Regulation, vol. 1–. New York: Academic Press, 1969–.

International Review of Cytology, vol. 1–. New York: Academic Press, 1952–.

8. Evolution

Evolutionary Biology, vol. 1–. New York: Appleton-Century-Crofts, 1964–.

9. Animal Physiology; Comparative Physiology; Comparative Anatomy (See also Biochemistry and Biophysics, and Cell and Molecular Biology; for Human Physiology see also Human Physiology and Medicine)

Advances in Comparative Physiology and Biochemistry, vol. 1–. New York: Academic Press, 1962–.

Advances in Insect Physiology, vol. 1–. London: Academic Press, 1963–.

Advances in Neurochemistry, vol. 1–. New York: Plenum Press, 1975–.

Annual Review of Physiology, vol. 1–. Palo Alto, Cal.: Annual Reviews, 1939–.

International Review of Neurobiology, vol. 1–. New York: Academic Press, 1959–.

Physiological Reviews, vol. 1–. Bethesda, Md.: American Physiological Society, 1921–.

Progress in Neurobiology, vol. 1–. Oxford, Eng.: Pergamon Press, 1973–.

10. Embryology; Development

Advances in Morphogenesis, vol. 1–. New York: Academic Press, 1961–.

Current Topics in Developmental Biology, vol. 1–. New York: Academic Press, 1966–.

11. Microbiology; Bacteriology; Virology (For Microscopy see also Biology, General)

Advances in Applied Microbiology, vol. 1–. New York: Academic Press, 1959–.

Advances in Immunology, vol. 1–. New York: Academic Press, 1961–.

Advances in Microbial Physiology, vol. 1–. New York: Academic Press, 1967–.

Advances in Virus Research, vol. 1–. New York: Academic Press, 1953–.

Annual Review of Microbiology, vol. 1–. Palo Alto, Cal.: Annual Reviews, 1947–.

Bacteriological Reviews, vol. 1–. Baltimore: American Society for Microbiology, 1937–.

CRC Critical Reviews in Microbiology, vol. 1–. Cleveland, Ohio: CRC Press, 1971–.

Modern Trends in Immunology, vol. 1–. New York: Appleton-Century-Crofts, 1963–.

12. Invertebrates

Advances in Insect Physiology, vol. 1–. London: Academic Press, 1963–.

Annual Review of Entomology, vol. 1–. Palo Alto, Cal.: Annual Reviews, 1956–.

13. Vertebrates; Herpatology; Icthyology; Mammalogy; Ornithology

International Zoo Yearbook, vol. 1–. London: Zoological Society, 1960–.

14. Parasitology

Advances in Parasitology, vol. 1–. London: Academic Press, 1963–.

15. Marine Biology (See also plant and animal headings for groups of marine species.)

Advances in Marine Biology, vol. 1–. London: Academic Press, 1963–.

Oceanography and Marine Biology, vol. 1–. London: Allen and Unwin, Ltd., 1963–.

16. Agriculture

Advances in Agronomy, vol. 1--. New York: Academic Press, 1949–.

17. Environmental Science; Pollution

Advances in Environmental Science and Technology, vol. 1–. New York: Wiley, 1969--.

CRC Critical Reviews in Environmental Control, vol. 1--. Cleveland, Ohio: CRC Press, 1970--.

18. Medicine

Advances in Drug Research, vol. 1–. New York: Academic Press, 1964--.

Advances in Pharmacology and Chemo-- therapy, vol. 1–. New York: Academic Press, 1962–.

Annual Reports in Medicinal Chemistry, vol. 1–. New York: Academic Press, 1965--.

Annual Review of Medicine, vol. 1--. Palo Alto, Cal.: Annual Reviews, 1950--.

Annual Review of Pharmacology and Toxi-- cology, vol. 1–. Palo Alto, Cal.: Annual Reviews, 1961–.

Nutrition Reviews, vol.1--. New York: Nutrition Foundation, Inc., 1942–.

Perspectives in Biology and Medicine, vol. 1–. Chicago: University of Chicago Press, 1957–.

Pharmacological Reviews, vol. 1--. Balti-- more: Williams & Wilkins Co., 1949–.

Progress in Medical Genetics, vol. 1--10. New York: Drune & Stratton, 1961–74.

Progress in Medical Genetics New Series, vol. 1–. New York: Saunders, 1976--.

Progress in Medical Virology, vol. 1–. New York: S. Karger, 1958--.

Progress in Medicinal Chemistry, vol. 1--. London: Butterworths, 1961--.

APPENDIX III

List of Guides to the Literature

1. Blake, John B. and C. Roos. *Medical Reference Works, 1679--1966: A Selected Bibliography and Supplements 1 and 2.* Chicago: Medical Library Association, 1967--1973. 3v.
2. Blanchard, Joy Richard and H. Ostvold. *Literature of Agricultural Research.* Berkeley: University of California Press, 1958. 231 pp.
3. Bottle, R.T. and H.V. Wyatt. *The Use of Biological Literature.* 2nd ed. Hamden, Conn.: Archon Books, 1971. 379 pp.
4. Brunn, Alice L. *How to Find Out in Pharmacy: A Guide to Sources of Pharmaceutical Information.* New York: Pergamon Press, 1969. 130 pp.
5. Jackson, Benjamin Daydon. *Guide to the Literature of Botany; . . .* New York: Hafner, 1964. 626 pp.
6. Jenkins, Frances Briggs. *Science Reference Sources.* 5th ed. Cambridge, Mass.: M.I.T. Press, 1969. 231 pp.
7. Kerker, Ann E. and H.T. Murphey. *Biological and Biomedical Resources Literature.* Lafayette, Ind.: Purdue University, 1968. 226 pp.
8. Lasworth, Earl James. *Reference Sources in Science and Technology.* Metuchen, N.J.: Scarecrow Press, 1972. 305 pp.
9. Malinowsky, Harold Robert, R.A. Gray, and D.A. Gray. *Sciences and Engineering Literature.* 2nd ed. Littleton, Colo.: Libraries Unlimited, 1976. 368 pp.
10. Mellon, Melvin Guy. *Chemical Publications, Their Nature and Use.* 4th ed. New York: McGraw--Hill, 1965. 324 pp.
11. Sheehy, Eugene. *Guide to Reference Books.* 9th ed. Chicago: American Library Association, 1976.
12. Smith, Roger Cletus and R.H. Painter. *Guide to the Literature of the Zoological Sciences.* 7th ed. Minneapolis: Burgess, 1966. 238 pp.
13. Smith, Roger Cletus and W.M. Reid. *Guide to the Literature of the Life Sciences.* 8th ed. Minneapolis: Burgess Publishing Co., 1972. 166 pp.
14. Walford, Albert John, ed. *Guide to Reference Material, vol. 1, Science and Technology.* 3rd ed. London: London Library Association, 1973. 615 pp.
15. Woodburn, Henry. *Using the Chemical Literature: a Practical Guide.* New York: M. Dekker, 1974. 302 pp.

APPENDIX IV

Basic Reference Sources for

Biology Courses

Note: The first library search with any of the following course--related bibliographies will be easier if used in conjunction with the main body of the book. To make it easier to locate materials, items are cited as they appear on Library of Congress catalog cards.

Symbols: * indicates a title described in the text, see the index for location. Each section of the outline is subdivided into the following reference tool categories:

A. Encyclopedias, Treatises, Texts, Dictionaries, Handbooks
B. Bibliographies
C. Review Serials
D. Data Compendia
E. Catalogs, Checklists, Field Guides
F. Methods
G. Indexes
H. Guides to the Literature

* * *

Outline

1. Biology, General
1B. Botany, General
1Z. Zoology, General
2. Plant Physiology (See also Botany, General)
3. Behavior
4. Genetics
5. Ecology (See also Marine Biology)
6. Biochemistry and Biophysics (See also Cell and Molecular Biology, Plant Physiology, and Animal Physiology)
7. Cell and Molecular Biology (See also Biochemistry and Biophysics, Plant Physiology, and Animal Physiology)
8. Evolution
9. Animal Physiology; Comparative Physiology; Comparative Anatomy (See also Biochemistry and Biophysics, and Cell and Molecular Biology; for Human Physiology see also Medicine)
10. Embryology; Development
11. Microbiology; Bacteriology; Virology (For Microscopy see also Biology, General)
12. Lower Plants; Algae; Fungi (Mycology); Mosses; Ferns
13. Higher Plants; Angiosperms; Gymnosperms
14. Invertebrates
15. Vertebrates; Herpatology; Icthyology; Mammalogy; Ornithology
16. Parasitology
17. Marine Biology (See also plant and animal headings for groups of marine species.)
18. Agriculture
19. Economic Botany
20. Environmental Science; Pollution
21. Medicine.

Bibliography

1. Biology, General

Style Manual

Council of Biology Editors. Committee on Form and Style. *CBE Style Manual.* 3rd ed. Washington, D.C.: American Institute of Biological Sciences, 1972. 297 pp.

A. Encyclopedias, Treatises, Texts, Dictionaries, Handbooks

Abercrombie, Michael, C.J. Hickman, and M.L. Johnson. *A Dictionary of Biology.* 6th ed. Baltimore: Penquin Books, 1973. 309 pp.

Britannica Yearbook of Science and the Future. Chicago: Encyclopaedia Britannica Inc.,1969--.

Clark, George Lindenberg. *The Encyclopedia of Microscopy.* New York: Reinhold, 1961. 693 pp.

Compton's Dictionary of the Natural Sciences. Chicago: Compton, 1966. 2 vols.

Encyclopaedia Britannica. Chicago: Encyclopaedia Britannica, 1976. 30 vols.

Fisher, Ronald Aylmer, and F. Yates. *Statistical Tables for Biological, Agricultural, and Medical Research.* 6th ed. New York: Hafner, 1963. 146 pp.

Gray, Peter. *Dictionary of Biology.* New York: Reinhold, 1967. 602 pp.

Gray, Peter. *The Encyclopedia of Microscopy and Microtechnique.* New York: Van Nostrand Reinhold, 1973. 638 pp.

Gray, Peter. *The Encyclopedia of the Biological Sciences.* 2nd ed. New York: Van Nostrand Reinhold, 1970. 1027 pp.

Henderson, Isabella Ferguson, and W.D. Henderson. *A Dictionary of Biological Terms.* 8th ed. Edinburgh & London: Oliver & Boyd, 1963. 640 pp.

McGraw--Hill Dictionary of the Life Sciences. New York: McGraw–Hill, 1976. 945 pp.

**McGraw--Hill Encyclopedia of Science and Technology.* 4th ed. New York: McGraw--Hill, 1977. 15 vols.

Palmer, Ephraim Laurence. *Fieldbook of Natural History.* 2nd ed. New York: McGraw--Hill, 1975. 779 pp.

Steen, Edwin B. *Dictionary of Biology.* New York: Barnes & Noble, 1971. 630 pp.

Van Nostrand's Scientific Encyclopedia. 4th ed. New York: Van Nostrand Reinhold, 1976. 2370 pp.

B. Bibliographies

Science for Society: a Bibliography. 1st ed.-- Washington: American Association for the Advancement of Science Commission on Science Education, 1970--.

C. Review Serials

American Scientist, vol. 1--. New Haven, Conn.: Society of the Sigma Xi, 1913--.

Biological Reviews, vol. 1--. Cambridge, University Press, 1923--.

Quarterly Review of Biology, vol. 1--. Stony Brook, New York: Stony Brook Foundation, Inc., 1926–.

Scientific American, vol. 1–. New York: Scientific American, 1845--.

D. Data Compendia

Altman, Philip L., and D.S. Ditmer, ed. *Biology Data Book.* 2nd ed. Bethesda: Federation of American Societies for Experimental Biology, 1972–74. 3 vols.

F. Methods

Grimstone, A.V. and R.J. Skaer. *A Guidebook to Microscopical Methods.* New York: Cambridge University Press, 1972. 134 pp.

Lillie, Ralph Dougall, and H.J. Conn. *Biological Stains: a Handbook . . .* Baltimore, Md.: Williams & Wilkins Co., 1976.

G. Indexes

**Biological Abstracts,* vol. 1–. Philadelphia: BioSciences Information Service, 1926--.

**Biological & Agricultural Index,* vol. 1--. New York: H.W. Wilson, 1916--.

**Biology Digest,* vol. 1–. Louisville, Ky.: Data Courier, 1974--.

**Bioresearch Index,* vol. 1–. Philadelphia: BioSciences Information Service, 1967--.

**Chemical Abstracts,* vol. 1–. Columbus, Ohio: American Chemical Society, 1907--.

**Current Contents, Agriculture, Biology & Environmental Sciences,* vol. 1--. Philadelphia: Institute for Scientific Information, 1970–.

**Current Contents, Physical & Chemical Sciences,* vol. 1--. Philadelphia: Institute for Scientific Information, 1961--.

**Science Citation Index,* vol. 1--. Philadelphia: Institute for Scientific Information, 1961–.

U.S. Superintendent of Documents. *Monthly Catalog of U.S. Government Publications,* vol. 1–. Washington, D.C.: U.S. Government Printing Office, 1895–.

H. Guides to the Literature

*Bottle, R.T., and H.V. Wyatt. *The Use of Biological Literature.* 2nd ed. Hamden, Conn.: Archon Books, 1971. 379 pp.

*Jenkins, Frances Briggs. *Science Reference Sources.* 5th ed. Cambridge, Mass.: M.I.T. Press, 1969. 231 pp.

*Kerker, Ann E., and H.T. Murphey. *Biological and Biomedical Resources Literature.* Lafayette, In.: Purdue University, 1968. 226 pp.

Lasworth, Earl James. *Reference Sources in Science and Technology.* Metuchen, N.J.: Scarecrow Press, 1972. 305 pp.

Malinowsky, Harold Robert, R.A. Gray, and D.A. Gray. *Sciences and Engineering Literature.* 2nd ed. Littleton, Colo.: Libraries Unlimited, 1976. 368 pp.

Smith, Roger Cletus, and W.M. Reid. *Guide to the Literature of the Life Sciences.* 8th ed. Minneapolis: Burgess Publishing Co., 1972. 166 pp.

*Walford, Albert John, ed. *Guide to Reference Material, vol. 1, Science and Technology.* 3rd ed. London: London Library Association, 1973. 615 pp.

1B. Botany, General

A. Encyclopedias, Treatises, Texts, Dictionaries, Handbooks

Bailey, Liberty Hyde. *The Standard Cyclopedia of Horticulture . . .* New York: Macmillan, 1914--1917. 3 vols.

Graf, Alfred Byrd. *Exotica 3; Pictorial Cyclopedia of Exotic Plants.* Century ed. Rutherford, N.J.: Roehrs Co., 1970. 1834 pp.

Jackson, Benjamin Daydon. *A Glossary of Botanic Terms.* London: Duckworth, 1928 (reprinted 1971). 481 pp.

Swartz, Delbert. *Collegiate Dictionary of Botany.* New York: Ronald Press, 1971. 520 pp.

Usher, George. *A Dictionary of Botany.* Prince--ton, N.J.: Van Nostrand Reinhold, 1966. 408 pp.

B. Bibliographies

HUNTIA: a Yearbook of Botanical and Horti--cultural Bibliography, vol. 1--. Pittsburgh, Pa.: Hunt Botanical Library, Carnegie Institute of Technology, 1964--.

Swift, Lloyd H. *Botanical Bibliographies . . .* Minneapolis: Burgess, 1970. 804 pp.

C. Review Serials

Advances in Botanical Research, vol. 1--. New York: Academic Press, 1963--.

Botanical Review, vol. 1--. New York: New York Botanical Garden, 1935--.

F. Methods

Klein, Richard M., and D.T. Klein. *Research Methods in Plant Science.* Garden City, New York: Natural History Press, 1970. 756 pp.

H. Guides to the Literature

Jackson, Benjamin Daydon. *Guide to the Litera--ture of Botany . . .* New York: Hafner, 1964. 626 pp.

1Z. Zoology, General

A. Encyclopedias, Treatises, Texts, Dictionaries, Handbooks (See also under Invertebrates and Vertebrates Sections)

Burton, Maurice, and R. Burton, eds. *The Inter--national Wildlife Encyclopedia.* New York: Marshall Cavendish Corp., 1969--1970. 20 vols.

**Grzimek's Animal Life Encyclopedia,* New York: Van Nostrand Reinhold, 1972–1976. 13 vols.

Leftwich, A.W. *A Dictionary of Zoology.* 3rd ed. New York: Crane, Russak, 1973. 478 pp.

Pennak, Robert William. *Collegiate Dictionary of Zoology.* New York: Ronald Press Co., 1964. 583 pp.

C. Review Serials

International Review of General and Experimental Zoology, vol. 1--. New York: Academic Press, 1964.

D. Data Compendia

CRC Handbook of Laboratory Animal Sciences. Cleveland, Ohio: CRC Press, 1974--1976. 3 vols.

G. Indexes

Wildlife Review, vol. 1--. Fort Collins, Colo.: Colorado State University, 1935--.

**The Zoological Record . . .* London: Zoological Society of London, 1865--.

H. Guides to the Literature

Smith, Roger Cletus, and R.H. Painter. *Guide to the Literature of the Zoological Sciences.* 7th ed. Minneapolis: Burgess, 1966. 238 pp.

2. Plant Physiology (See also Botany, General)

A. Encyclopedias, Treatises, Texts, Dictionaries, Handbooks

Handbuch der Pflanzenphysiologie: Encyclopedia of Plant Physiology. Berlin: Springer, 1955--67 18 vols.

Steward, Frederick Campion, ed. *Plant Physi--ology: A Treatise.* New York: Academic Press, 1960--. (11 vols. in 1976).

B. Bibliographies

Barton, Lela Viola. *Bibliography of Seeds.* New York: Columbia University Press, 1967. 858 pp.

Miller, Robert Harold. *Root Anatomy and Mor--phology: A Guide to the Literature.* Hamden, Conn.: Archon Books, 1974. 271 pp.

C. Review Serials

Annual Review of Phytopathology, vol. 1--. Palo Alto, Cal.: Annual Reviews, 1963--.

Annual Review of Plant Physiology, vol. 1–. Palo Alto, Cal.: Annual Reviews, 1950–.

3. Behavior (Including Psychology)

A. Encyclopedias, Treatises, Texts, Dictionaries, Handbooks

Drever, James. *A Dictionary of Psychology.* Rev. ed. Baltimore: Penguin Books, 1964. 320 pp.
Encyclopedia of Psychology. New York: Herder & Herder, 1972. 3 vols.
Goldenson, Robert M. *The Encyclopedia of Human Behavior, Psychology, Psychiatry, and Mental Health.* Abridged and revised. New York: Dell, 1975. 892 pp.
Hinde, Robert A. *Animal Behaviour; a Synthesis of Ethology and Comparative Psychology.* 2nd ed. New York: McGraw–Hill, 1970. 876 pp.

C. Review Serials

Advances in Psychobiology, vol. 1–. New York: Wiley–Interscience, 1972–.
Advances in the Study of Behavior, vol. 1–. New York: Academic Press, 1965–.
Annual Review of Psychology, vol. 1–. Palo Alto, Cal.: Annual Reviews, 1951–.

G. Indexes

Psychological Abstracts, vol. 1–. Lancaster, Pa.: American Psychological Association, 1927–.

4. Genetics

A. Encyclopedias, Treatises, Texts, Dictionaries, Handbooks

King, Robert C. *A Dictionary of Genetics.* 2nd ed. rev. New York: Oxford, 1974. 375 pp.
Rieger, R., and others. *A Glossary of Genetics and Cytogenetics, Classical and Molecular.* 4th completely rev. ed. Berlin: Springer–Verlag, 1976. 647 pp.
Srb, Adrian M., R.D. Owen, and R.E. Edgar. *General Genetics.* 2nd ed. San Francisco: W.H. Freeman, 1965. 557 pp.

B. Bibliographies

McKusick, Victor A. *Mendelian Inheritance in Man; Catalogs of Autosomal Dominant, Autosomal Recessive, and X–Linked Phenotypes.* 3rd ed. Baltimore: Johns Hopkins Press, 1975. 837 pp.
Muller, Hermann Joseph. *Bibliography on the Genetics of Drosophila.* Edinburgh: Oliver and Boyd, 1939–. (6 vols. in 1976.)

C. Review Serials

Advances in Genetics, vol. 1–. New York: Academic Press, 1947–.
Advances in Human Genetics, vol. 1–. New York: Plenum, 1970–.
Annual Review of Genetics, vol. 1–. Palo Alto, Cal.: Annual Reviews, 1967–.

G. Indexes

Current Advances in Genetics; a Monthly Subject Categorised Listing of Titles in Genetics Compiled from the Current Literature, vol. 1–. Oxford, Eng.: Pergamon Press, 1976–.
Genetics Abstracts, vol. 1–. London: Information Retrieval, Ltd., 1968–.

5. Ecology (See also Marine Biology)

A. Encyclopedias, Treatises, Texts, Dictionaries, Handbooks

Carpenter, John Richard. *An Ecological Glossary.* New York: Hafner, 1956 (c. 1938). 306 pp.
Hanson, Herbert Christian. *Dictionary of Ecology.* New York: Philosophical Library, 1962. 382 pp.
Odum, Eugene Pleasants. *Fundamentals of Ecology.* 3rd ed. Philadelphia: Saunders, 1971. 574 pp.
Patten, Bernard C., ed. *Systems Analysis and Simulation in Ecology.* New York: Academic Press, 1971–. (4 vols. as of 1976).
Ricklefs, Robert E. *Ecology.* Newton, Mass.: Chiron Press, 1973. 861 pp.

C. Review Serials

**Advances in Ecological Research,* vol. 1–. New York: Academic Press, 1962–.
Annual Review of Ecology and Systematics, vol. 1–. Palo Alto, Cal.: Annual Reviews, 1970–.

D. Data Compendia

Altman, Philip L., ed. *Environmental Biology.* Bethesda: Federation of American Societies for Experimental Biology, 1966. 694 pp.

G. Indexes

Current Advances in Ecological Sciences; a Monthly Subject Categories Listing of Titles in Ecological Sciences Compiled from the Current Literature, vol. 1--. Oxford, Eng.: Pergamon Press, 1975--.

Ecological Abstracts, vol. 1--. Norwich, Eng.: Geo Abstracts Lt., 1974--.

Geo Abstracts B. Biogeography and Climatology, vol. 1--. Norwich, Eng.: University of East Anglia, 1966--.

6. Biochemistry and Biophysics (See also Cell and Molecular Biology, Plant Physiology, and Animal Physiology)

A. Encyclopedias, Treatises, Texts, Dictionaries, Handbooks

Barman, Thomas E. *Enzyme Handbook.* Berlin: Springer-Verlag, 1969--1974. 3 vols.

Boyer, Paul D., ed. *The Enzymes.* 3rd ed. New York & London: Academic Press, 1970--. (12 vols. as of 1976).

Commission on Biochemical Nomenclature. *Enzyme Nomenclature.* Amsterdam: Elsevier, 1973. 449 pp.

Florkin, Marcel, and H.S. Mason. *Comparative Biochemistry: a Comprehensive Treatise.* New York: Academic Press, 1960--64. 7 vols.

Florkin, Marcel, and E.H. Stotz, eds. *Comprehensive Biochemistry.* Amsterdam: Elsevier, 1962--. (32 vols. as of 1976).

Greenberg, David Morris. *Metabolic Pathways.* 3rd ed. New York: Academic Press, 1967--72. 6 vols.

Lehninger, Albert L. *Biochemistry: The Molecular Basis of Cell Structure and Function.* 2nd ed. New York: Worth Publishers, 1975. 1104 pp.

Neurath, Hans, R.L. Hill, ed. *The Proteins: Composition, Structure, and Fuction.* 2nd and 3rd eds. New York: Academic Press, 2nd ed.: 1963--70, 5 vols., 3rd ed.: 1975--. (2 vols. as of 1976).

Pincus, Gregory, and K.V. Thimann, ed. *The Hormones: Physiology, Chemistry, and Applications.* New York: Academic Press, 1948--1964. 5 vols.

Sebrell, William Henry, and R.S. Harris, ed. *The Vitamins: Chemistry, Physiology, Pathology, Methods.* 2nd ed. New York: Academic Press, 1967--1972. 7 vols.

Williams, R.J., and E.M. Lansford, eds. *The Encyclopedia of Biochemistry.* New York: Reinhold, 1967. 876 pp.

Windholz, Martha, ed. *The Merck Index; an Encyclopedia of Chemicals and Drugs.* 9th ed. Rahway, N.J.: Merck, 1976. 1937 pp.

Zweig, Gunter, and J. Sherman. *Handbook of Chromatography.* Cleveland, Ohio: CRC Press, 1972. 2 vols.

B. Bibliographies

Semenuk, Nick S., and H. Zimmenberg. *Cyclic AMP, a Classified Bibliography of Publications.* New Brunswick, N.J.: E.R. Squibb, 1970--. (6 vols. as of 1976).

C. Review Serials

Advances in Biological and Medical Physics, vol. 1--. New York: Academic Press, 1948--.

Advances in Biophysics, vol. 1--. Tokyo: University of Tokyo Press, 1970--.

Advances in Carbohydrate Chemistry and Biochemistry, vol. 1--. New York: Academic Press, 1945--.

Advances in Enzymology and Related Areas of Molecular Biology, vol. 1--. New York: Interscience, Wiley, 1941--.

Advances in Lipid Research, vol. 1--. New York: Academic Press, 1963--.

Advances in Microbial Physiology, vol. 1--. New York: Academic Press, 1967--.

Advances in Protein Chemistry, vol. 1--. New York: Academic Press, 1944--.

Advances in Radiation Biology, vol. 1--. New York: Academic Press, 1964--.

Annual Review of Biochemistry, vol. 1--. Palo Alto, Cal.: Annual Reviews, 1932--

Annual Review of Phytopathology, vol. 1--. Palo Alto, Cal.: Annual Reviews, 1963--.

Critical Reviews in Biochemistry, vol. 1--. Cleveland, Ohio: CRC Press, 1972--.

Essays in Biochemistry, vol. 1--. New York: Academic Press, 1965--.

Progress in Biophysics and Molecular Biology, vol. 1--. Oxford, Eng.: Pergamon Press, 1950--.

Progress in Nucleic Acid Research and Molecular Biology, vol. 1--. New York: Academic Press, 1963--.

Quarterly Reviews of Biophysics, vol. 1--. London: Cambridge University Press, 1968--.

Vitamins and Hormones, Advances in Research and Applications, vol. 1--. New York: Academic Press, 1943--.

D. Data Compendia

Damm, Henry Clarence, ed. *The Handbook of Biochemistry and Biophysics.* Cleveland, Ohio: World Publishing Co., 1966. 736 pp.

Data for Biochemical Research. 2nd ed. Oxford: Clarendon Press, 1969. 654 pp.

Handbook of Biochemistry and Molecular Biology. 3rd ed. Cleveland, Ohio: CRC Press, 1976. 8 vols.

Handbook of Chemistry and Physics; a Ready-- Reference Book of Chemical and Physical Data. 1st-- ed. Cleveland, Ohio: CRC Press, 1913--. Annual.

F. Methods

Colowick, S.P., and N.O. Kaplan, eds. *Methods in Enzymology.* New York: Academic Press, 1955--. vol. 1--.

Damm, Henry Clarence, ed. *Methods and Ref-- erences in Biochemistry and Biophysics.* Cleve-- land, Ohio: World Publishing Co., 1966. 1 vol. (loose-leaf).

Methods of Biochemical Analysis, vol. 1--. New York: Interscience Publishers, 1954--.

G. Indexes

**Chemical Abstracts,* vol. 1--. Columbus, Ohio: American Chemical Society, 1907--.

H. Guides to the Literature

Mellon, Melvin Guy. *Chemical Publications, Their Nature and Use.* 4th ed. New York: McGraw-- Hill, 1965. 324 pp.

Woodburn, Henry. *Using the Chemical Literature: a Practical Guide.* New York: M. Dekker, 1974. 302 pp.

7. Cell and Molecular Biology (See also Bio-- chemistry and Biophysics, Plant Physiology, and Animal Physiology)

A. Encyclopedias, Treatises, Texts, Dictionaries, Handbooks

Brachet, Jean, ed., and A.F. Mirsky. *The Cell; Biochemistry, Physiology, Morphology.* New York: Academic Press, 1959--1964. 6 vols.

Evans, Anthony. *Glossary of Molecular Biology.* New York: Wiley, 1975, 55 pp.

C. Review Serials

Advances in Cell and Molecular Biology, vol. 1--. New York: Academic Press, 1971--.

Advances in Cell Biology, vol. 1--. Englewood Cliffs, N.J.: Prentice--Hall, 1970--.

Current Topics in Cellular Regulation, vol. 1--. New York: Academic Press, 1969--.

International Review of Cytology, vol. 1--. New York: Academic Press, 1952--.

8. Evolution

A. Encyclopedias, Treatises, Texts, Dictionaries, Handbooks

Ehrlich, Paul R. *The Process of Evolution.* 2nd ed. New York: McGraw-Hill, 1974. 378 pp.

Grzimek, Bernhard. *Grzimek's Encyclopedia of Evolution.* New York: Van Nostrand Reinhold, 1976. 560 pp.

Mayr, Ernst. *Animal Species and Evolution.* Cambridge: Harvard University Press, 1963. 797 pp.

Salthe, Stanley N. *Evolutionary Biology.* New York: Holt, Rinehart and Winston, 1972. 437 pp.

B. Bibliographies

Fay, George Emory. *A Bibliography of Fossil Man,* vol. 1--. Magnolia, Ark.: Dept. of Sociol-- ogy and Anthropology, Southern State College, 1959--. (2 vols. & 2 supps. as of 1976).

C. Review Serials

Evolutionary Biology, vol. 1--. New York: Apple-- ton--Century--Crofts, 1964--.

9. Animal Physiology; Comparative Physiology; Comparative Anatomy (See also Biochemistry and Biophysics, and Cell and Molecular Biology; for Human Physiology see also Human Physiology and Medicine)

A. Encyclopedias, Treatises, Texts, Dictionaries, Handbooks

Anthony, Catherine P. *Textbook of Anatomy and Physiology.* 9th ed. St. Louis: Mosby, 1975. 598 pp.

Bell, George H. *Textbook of Physiology and Bio-- chemistry.* 8th ed. Baltimore: Williams & Wilkins. 1160 pp.

Handbook of Physiology; a Critical, Comprehen-- sive Presentation of Physiological Knowledge and Concepts. Washington, D.C.: American Physiological Society, 1959--. (23 vols. as of 1976).

Hoar, William Stewart, and D.J. Randall, ed. *Fish Physiology.* New York: Academic Press, 1969--1971. 6 vols.

Hoar, William Stewart. *General and Compara-- tive Physiology.* 2nd ed. Englewood Cliffs,

N.J.: Prentice--Hall, 1975. 848 pp.

Kent, George Cantine. *Comparative Anatomy of the Vertebrates.* 3rd ed. St. Louis: Mosby, 1973. 414 pp.

Prosser, Clifford Ladd, ed. *Comparative Animal Physiology.* 3rd ed. Philadelphia: Saunders, 1973. 966 pp.

Rockstein, Morris, ed. *The Physiology of Insecta.* 2nd ed. New York: Academic Press, 1973--1974. 6 vols.

Turner, Clarence Donnell, and J.T. Bagnara. *General Endocrinology.* 6th ed. Philadelphia: Saunders, 1976. 596 pp.

B. Bibliographies

Bibliography of Reproduction; a Classified Monthly Title List, vol. 1–. Cambridge, Eng.: Reproduction Research Information Service, 1963--.

C. Review Serials

Advances in Comparative Physiology and Biochemistry, vol. 1--. New York: Academic Press, 1962–.

Advances in Insect Physiology, vol. 1--. London: Academic Press, 1963--.

Advances in Neurochemistry, vol. 1--. New York: Plenum Press, 1975--.

Annual Review of Physiology, vol. 1–. Palo Alto: Cal.: Annual Reviews, 1939--.

International Review of Neurobiology, vol. 1–. New York: Academic Press, 1959–.

Physiological Reviews, vol. 1–. Bethesda: American Physiological Society, 1921–.

Progress in Neurobiology, vol. 1–. Oxford, Eng.: Pergamon Press, 1973--.

D. Data Compendia

Altman, Philip L. *Blood and Other Body Fluids.* Washington, D.C.: Federation of American Societies for Experimental Biology, 1961. 540 pp.

Altman, Philip L., and D.S. Dittmer. *Metabolism.* Bethesda: Federation of American Societies for Experimental Biology, 1968. 737 pp.

International Commission on Radiological Protection. Committee II on Permissable Dose for Internal Radiation. Task Group on Reference Man. *Report of the Task Group on Reference Man.* Oxford, Eng.: Pergamon Press, 1975. 480 pp.

Respiration and Circulation. Bethesda: Federation of American Societies for Experimental Biology, 1971. 930 pp.

10. Embryology; Development

A. Encyclopedias, Treatises, Texts, Dictionaries, Handbooks

Balinsky, Boris I. *Introduction to Embryology.* 4th ed. Philadelphia: Saunders, 1975. 648 pp.

Berrill, Norman John, and G. Karp. *Development.* New York: McGraw–Hill, 1976. 566 pp.

Romanoff, Alexis. *The Avian Embryo; Structural and Functional Development.* New York: Macmillan, 1960. 1305 pp.

Rugh, Roberts. *Vertebrate Embryology: the Dynamics of Development.* New York: Harcourt, Brace & World, 1964. 600 pp.

C. Review Serials

Advances in Morphogenesis, vol. 1--. New York: Academic Press, 1961–.

Current Topics in Developmental Biology, vol. 1--. New York: Academic Press, 1966--.

D. Data Compendia

Altman, Philip L., and D.S. Dittmer. *Growth Including Reproduction and Morphological Development.* Washington, D.C.: Federation of American Societies for Experimental Biology, 1962. 608 pp.

11. Microbiology; Bacteriology; Virology (For Microscopy see also Biology, General)

A. Encyclopedias, Treatises, Texts, Dictionaries, Handbooks

Davis, Bernard David. *Microbiology; Including Immunology and Molecular Genetics.* 2nd ed. Hagerstown, Md.: Harper & Row, 1973. 1562 pp.

Fraenkel--Conrat, Heinz, and R.R. Wagner, eds. *Comprehensive Virology.* New York: Plenum Press, 1974--. (7 vols. as of 1976).

Herbert, William John, and P.C. Wilkinson, eds. *A Dictionary of Immunology.* Oxford, Eng.: Blackwell Scientific Publications, 1971. 195 pp.

Rose, Anthony H., and J.S. Harrison, ed. *The Yeasts.* London: Academic Press, 1969--1971. 3 vols.

Topley, William Whiteman Carlton. *Topley and Wilson's Principles of Bacteriology and Immunity.* 6th ed. London: E. Arnold, 1975. 2 vols.

C. Review Serials

Advances in Applied Microbiology, vol. 1--. New York: Academic Press, 1959--.

Advances in Immunology, vol. 1--. New York: Academic Press, 1961--.

Advances in Microbial Physiology, vol. 1--. New York: Academic Press, 1967--.

Advances in Virus Research, vol. 1--. New York: Academic Press, 1953--.

Annual Review of Microbiology, vol. 1--. Palo Alto, Cal.: Annual Reviews, 1947--.

Bacteriological Reviews, vol. 1--. Baltimore: American Society for Microbiology, 1937--.

CRC Critical Reviews in Microbiology, vol. 1--. Cleveland, Ohio: CRC Press, 1971--.

Modern Trends in Immunology, vol. 1--. New York: Appleton--Century--Crofts, 1963--.

D. Data Compendia

CRC Handbook of Microbiology. Cleveland Ohio: CRC Press, 1973--1974. 4 vols.

D. Catalogs, Checklists, Field Guides

Bergey's Manual of Determinative Bacteriology. 8th ed. Baltimore: Williams & Wilkins, 1974. 1268 pp.

Cowan and Steel's Manual for the Identification of Medical Bacteria. 2nd ed. New York: Cambridge University Press, 1974. 238 pp.

Index Bergeyana: an Annotated Alphabetical Listing of Names of the Taxa of the Bacteria. 7th ed. Baltimore: Williams & Wilkins, 1966. 1472 pp.

Skerman, V.B.D. *A Guide to the Identification of the Genera of Bacteria.* 2nd ed. Baltimore: Williams & Wilkins, 1973. 303 pp.

F. Methods

Digestive Ferments Company. Difco Laboratories, Detroit. *Difco Manual of Dehydrated Culture Media and Reagents for Microbiological and Clinical Laboratory Procedures.* 9th ed. Detroit, Mich.: Difco Laboratories, 1953. 350 pp.

Gibbs, Brian and F.A. Skinner, eds. *Identification Methods for Microbiologists.* London: Academic Press, 1966--1968. 2 vols.

Norris, John Robert, and D.W. Ribbons, eds. *Methods in Microbiology.* London: Academic Press, 1969--1973. 8 vols.

G. Indexes

Microbiology Abstracts, vol. 1--. London: Information Retrieval, Ltd., 1966--.

Virology Abstracts, vol. 1--. London: Information Retrieval, Ltd., 1967--.

12. Lower Plants; Algae; Fungi (Mycology); Mosses; Ferns

A. Encylopedias, Treatises, Texts, Dictionaries, Handbooks

Ainsworth, Geoffrey Clough. *Ainsworth & Bisby's Dictionary of the Fungi.* 6th ed. Kew, Surrey: Commonwealth Mycological Institute, 1971. 663 pp.

Ainsworth, Geoffrey Clough, and A.S. Sussman, eds. *The Fungi; an Advanced Treatise.* New York: Academic Press, 1965--1973. 5 vols.

Fritsch, Felix Eugene. *The Structure and Reproduction of the Algae.* Cambridge, Eng.: Cambridge University Press, 1935--1945. 2 vols.

International Code of Botanical Nomenclature, adopted by the Ninth International Botanical Congress, Seattle, August 1969. Utrecht: Oosthoek, 1972. 426 pp.

Mycological Society of America. Mycological Guidebook Committee. *Mycology Guidebook.* Seattle: University of Washington Press, 1974. 703 pp.

Snell, Walter Henry, and E.A. Dick. *A Glossary of Mycology.* Rev. ed. Cambridge: Harvard University Press, 1971. 181 pp.

Taylor, William R. *Marine Algae of the Northeastern Coast of North America.* Rev. ed. Ann Arbor: University of Michigan Press, 1957. 509 pp.

Willis, John Christopher. *A Dictionary of the Flowering Plants and Ferns.* 8th ed. Cambridge: Cambridge University Press, 1973. 1245 pp.

Wit, H.C.D., ed. *Plants of the World.* New York: Dutton, 1966--1969. 3 vols.

B. Bibliographies

Bachmann, Barbara J. and W.N. Strickland. *Neurospora Bibliography and Index.* New Haven, Conn.: Yale University Press, 1965. 225 pp.

Blake, Sidney Fay. *Geographical Guide to Floras of the World, an Annotated List With Special Reference to Useful Plants and Common Plant Names.* Washington, D.C.: Government Printing Office, 1942. 2 vols.

Solberg, Otto Thomas, and T.W. Gadella. *Biosystematic Literature: Contributions to a Biosystematic Literature Index (1945--1964).* Utrecht: International Association for Plant Taxonomy, 1970. 566 pp.

Stafleu, Frans Antonie. *Taxonomic Literature: a Selective Guide to Botanical Publications.* Utrecht: International Bureau for Plant Taxonomy and Nomenclature, 1967. 556 pp.

E. Catalogs, Checklists, Field Guides

Barnett, Horace Leslie, and B.B. Hunter. *Illustrated Genera of Imperfect Fungi.* 3rd ed. Minneapolis: Burgess, 1972. 241 pp.

Bower, Frederick Orpen. *The Ferns (Filicales) Treated Comparatively With a View to Their Natural Classification.* Cambridge: Cambridge University Press, 1923–1928. 3 vols.

Edmondson, Walles Thomas, ed. *Fresh--Water Biology.* 2nd ed. New York: Wiley, 1959. 1248 pp.

Grout, Abel Joel. *Moss Flora of North America, North of Mexico.* New York: The author, 1928. 3 vols.

Rinaldi, Augusto, and V. Tyndalo. *The Complete Book of Mushrooms . . .* New York: Crown, 1974. 331 pp.

Schuster, Rudolf. *The Hepaticae and Anthocerotae of North America East of the Hundreth Meridian.* New York: Columbia University Press, 1966--1976. 3 vols.

Smith, Gilbert Morgan. *The Fresh--Water Algae of the United States.* 2nd ed. New York: McGraw-Hill, 1950. 719 pp.

For a fuller list of keys and other taxonomic works see Roger C. Smith, *Guide to the Literature of the Life Sciences.* Minneapolis: Burgess, 1972.

F. Methods

Stein, Janet R., ed. *Handbook of Phycological Methods: Culture Methods and Growth Measurements.* New York: Cambridge University Press, 1973. 448 pp.

G. Indexes

Abstracts of Mycology, vol. 1--. Philadelphia: BioSciences Information Service of Biological Abstracts, 1967--.

13. Higher plants; Angiosperms; Gymnosperms

A. Encyclopedias, Treatises, Texts, Dictionaries, Handbooks

Carleton, R. Milton. *Index to Common Names of Herbaceous Plants.* Boston: G.K. Hall, 1959. 129 pp.

Featherly, Henry Ira. *Taxonomic Terminology of the Higher Plants.* Ames, Iowa: Iowa State College Press, 1954. 166 pp.

International Code of Botanical Nomenclature, adopted by the Ninth International Botanical Congress, Seattle, August 1969. Utrecht: Oosthoek, 1972. 426 pp.

Kingsbury, John Merriam. *Poisonous Plants of the United States and Canada.* Englewood Cliffs, N.J.: Prentice--Hall, 1964. 626 pp.

Lawrence, George Hill Mathewson. *Taxonomy of Vascular Plants.* New York: Macmillan, 1951. 823 pp.

Willis, John Christopher. *A Dictionary of the Flowering Plants and Ferns.* 8th ed. rev. by H.K. Airy Shaw. Cambridge: Cambridge University Press, 1973. 1245 pp.

Wit, H.C.D., ed. *Plants of the World.* New York: Dutton, 1966–1969. 3 vols.

B. Bibliographies

Barton, Lela Viola. *Bibliography of Seeds.* New York: Columbia University Press, 1967. 858 pp.

Blake, Sidney Fay. *Geographical Guide to Floras of the World, an Annotated List With Special Reference to Useful Plants and Common Plant Names.* Washington, D.C.: Government Printing Office, 1942. 2 vols.

Index Kewensis; Index Kewensis: an Enumeration of the Genera and Species of Flowering Plants from Time of Linnaeus to the Year 1885 Inclusive . . . Oxford, Eng.: Clarendon, 1893--1895. 2 vols.

-----. *Supplementum,* vol. 1--. Oxford, Eng.: Clarendon, 1901–. (15 vols. as of 1976).

Solbrig, Otto Thomas, and T.W.J. Gadella. *Biosystematic Literature: Contributions to a Biosystematic Literature Index (1945--1964).* Utrecht: International Association for Plant Taxonomy, 1970. 566 pp.

Stafleu, Frans Antonie. *Taxonomic Literature: a Selective Guide to Botanical Publications, with Dates, Commentaries and Types.* Utrecht: International Bureau for Plant Taxonomy and Nomenclature, 1967. 556 pp.

E. Catalogs, Checklists, Field Guides

Abrams, Le Roy. *An Illustrated Flora of the Pacific States: Washington, Oregon, and California.* Stanford, Cal.: Stanford University Press, 1923--1960. 4 vols.

Bailey, Liberty Hyde. *Manual of Cultivated Plants Most Commonly Grown in the Continental United States and Canada.* Rev. ed. New York: Macmillan, 1949. 1116 pp.

Gray, Asa. *Gray's Manual of Botany. A Handbook of the Flowering Plants and Ferns of the Central and Northeastern United States and Adjacent Canada.* 8th ed. New York: American Book Co., 1950. 1632 pp.

Hutchinson, John. *The Families of Flowering Plants.* 3rd ed. Oxford, Eng.: Clarendon Press, 1973. 968 pp.

Hutchinson, John. *The Genera of Flowering Plants, Angiospermae.* Oxford, Eng.: Clarendon Press, 1964--. (2 vols. as of 1976).

Hutchinson, John. *Key to the Families of Flower--ing Plants of the World.* Revised and enlarged for use as a supplement to *The Genera of Flowering Plants.* (q.v.). Oxford, Eng.: Claren--don Press, 1967. 117 pp.

Petrides, George A. *A Field Guide to Trees and Shrubs: Field Marks of All Trees, Shrubs, and Woody Vines that Grow Wild in the North--eastern and North--central United States and in Southeastern and South--central Canada.* 2nd ed. Boston: Houghton Mifflin, 1972. 428 pp.

Preson, Richard J. *North American Trees (Exclu--sive of Mexico and Topical United States).* 2nd ed. Ames, Iowa: Iowa State University Press, 1961. 395 pp.

Rehder, Alfred. *Manual of Cultivated Trees and Shrubs Hardy in North America, Exclusive of the Subtropical and Warmer Temperate Regions.* 2nd ed. New York: Macmillan, 1940. 996 pp.

Rickett, Harold William. *Wild Flowers of the United States.* New York: McGraw-Hill, 1966--1975. 15 vols.

Small, John Kunkel. *Manual of the Southeastern Flora: Being Descriptions of the Seed Plants Growing Naturally in Florida, Alabama, Missis--sippi, Eastern Louisiana, Tennessee, North Carolina, South Carolina and Georgia.* Chapel Hill, N.C.: University of North Carolina, 1933. 1554 pp.

Weber, William A. *Rocky Mountain Flora: A Field Guide for the Identification of the Ferns, Conifers, and Flowering Plants of the Southern Rocky Mountains.* 4th ed. Boulder, Col.: Colorado Associated University Press, 1972. 438 pp.

For a fuller list of keys and other taxonomic works see Roger C. Smith, *Guide to the Literature of the Life Sciences.* Minneapolis: Burgess, 1972.

14. Invertebrates

A. Encyclopedias, Treatises, Texts, Dictionaries, Handbooks

Caras, Roger. *Venomous Animals of the World.* Englewood Cliffs, N.J.: Prentice--Hall, 1974. 362 pp.

Chapman, Reginald Frederick. *The Insects: Structure and Function.* 2nd ed. New York: American Elsevier Pub. Co., 1971. 819 pp.

De la Torre--Bueno, Jose Rollin. *A Glossary of Entomology.* Brooklyn, N.Y.: Brooklyn Entomological Society, 1950. 336 pp.

*Grzimek, Bernard, ed. *Grzimek's Animal Life Encyclopedia: Vol. 1, Lower Animals; Vol. 2, Insects; Vol. 3, Mollusks and Echinoderms.* New York: Van Nostrand Reinhold, 1974--75.

Hyman, Libbie H. *The Invertebrates.* New York: McGraw--Hill, 1940-1967. 6 vols.

International Commission on Zoological Nomen-clature. *International Code of Zoological Nomenclature.* London: International Trust for Zoological Nomenclature, 1964. 176 pp.

Kaestner, Alfred. *Invertebrate Zoology.* New York: Wiley, 1967--. (3 vols. as of 1976).

Kudo, Richard Roksabro. *Protozoology . . .* 5th ed. Springfield, Ill.: Thomas, 1966. 1174 pp.

Larousee Encyclopedia of Animal Life. New York: McGraw--Hill, 1967. 640 pp.

Leftwich, A.W. *A Dictionary of Entomology.* New York: Crane Russak, 1976. 360 pp.

Metcalf, Clell Lee, and W.P. Flint. *Destructive and Useful Insects; Their Habits and Control.* 4th ed. New York: McGraw--Hill, 1962. 1087 pp.

Rockstein, Morris, ed. *The Physiology of Insecta.* 2nd ed. New York: Academic Press, 1973--. (6 vols. as of 1976).

Smart, Paul. *The International Butterfly Book.* New York: Crowell, 1975. 275 pp.

Swan, Lester A., and C.S. Papp. *The Common Insects of North America.* New York: Harper & Row, 1972. 750 pp.

Watson, Allan, and P.E.S. Whalley. *The Dictionary of Butterflies and Moths in Color.* New York: McGraw-Hill, 1975. 296 pp.

Wigglesworth, Vincent Brian. *The Principles of Insect Physiology.* 7th ed. London: Chapman and Hall, 1972. 827 pp.

B. Bibliographies

Chamberlin, Willard Joseph. *Entomological Nomenclature and Literature.* 3rd ed. Du--buque, Iowa: W.C. Brown, 1952. 141 pp.

Muller, Hermann Joseph. *Bibliography on the Genetics of Drosophila.* Edinburgh: Oliver and Boyd; Bloomington, Ind.: Indiana University, 1939--. (6 vols. as of 1976).

C. Review Serials

Advances in Insect Physiology, vol. 1--. London: Academic Press, 1963--.

Annual Review of Entomology, vol. 1--. Palo Alto, Cal.: Annual Reviews, 1956--.

E. Catalogs, Checklists, Field Guides

Borror, Donald Joyce, and R.E. White. *A Field Guide to the Insects of America North of Mexico.* Boston: Houghton Mifflin, 1970. 404 pp.

Edmondson, Walles, ed. *Fresh--water Biology.* 2nd ed. New York: Wiley, 1959. 1248 pp.

Gosner, Kenneth L. *Guide to Identification of Marine and Estuarine Invertebrates, Cape Hatteras to the Bay of Fundy.* New York: Wiley, 1971. 693 pp.

Morris, Percy A. *A Field Guide to Shells of the Atlantic and Gulf Coasts and the West Indies.* 3rd ed. Boston: Houghton Mifflin, 1973. 330 pp.

Morris, Percy A. *A Field Guide to Shells of the Pacific Coast and Hawaii.* 2nd ed. Boston: Houghton Mifflin, 1974. 220 pp.

Neave, Sheffield A., ed. *Nomenclator Zoologicus: a List of Names of Genera and Subgenera in Zoology.* London: Zoological Society of London, 1939--1940. 4 vols.

------. *Supplement,* vol. 1--, 1950--. (2 vols. as of 1976).

For a fuller list of keys and other taxonomic works see Roger C. Smith, *Guide to the Litera-- ture of the Life Sciences.* Minneapolis: Burgess, 1972.

G. Indexes

Abstracts of Entomology, vol. 1--. Philadelphia: BioSciences Information Service of Biological Abstracts, 1970--.

Entomology Abstracts, vol. 1--. London: Information Retrieval Ltd., 1969--.

Review of Applied Entomology, vol. 1--. Farnham Royal, Eng.: Commonwealth Agricultural Bureaux, 1913--.

15. Vertebrates; Herpatology; Icthyology; Mam-- malogy; Ornithology

A. Encyclopedias, Treatises, Texts, Dictionaries, Handbooks

General

Caras, Roger. *Venomous Animals of the World.* Englewood Cliffs, N.J.: Prentice--Hall, 1974. 362 pp.

International Commission on Zoological Nomen-- clature. *International Code of Zoological Nomenclature.* London: International Trust for Zoological Nomenclature, 1964. 176 pp.

Larousse Encyclopedia of Animal Life. New York: McGraw--Hill, 1967. 640 pp.

Young, John Z. *The Life of Vertebrates.* 2nd ed. Oxford, Eng.: Clarendon Press, 1962. 820 pp.

Fish

**Grzimek's Animal Life Encyclopedia: Vol. 4, Fishes I; Vol. 5, Fishes II and Amphibians.* New York: Van Nostrand Reinhold, 1972-- 1975.

Herald, Earl S. *Living Fishes of the World.* Garden City, New York: Doubleday, 1962. 303 pp.

Nikol'skii, Georgii V. *The Ecology of Fishes.* London: Academic Press, 1963. 352 pp.

Sterba, Gunther. *Freshwater Fishes of the World.* New York: Viking Press, 1962. 878 pp.

Reptiles and Amphibians

Carr, Archie Fairly. *Handbook of Turtles; the Turtles of the United States, Canada, and Baja California.* Ithaca, N.Y.: Comstock Pub. Associates, 1952. 542 pp.

Cochran, Doris M. *Living Amphibians of the World.* Garden City, N.Y.: Doubleday, 1961. 199 pp.

Ditmars, Raymond Lee. *The Reptiles of North America: a Review of the Crocodilians, Lizards, Snakes, Turtles and Tortoises Inhabiting the United States and Northern Mexico.* Garden City, N.Y.: Doubleday, 1936. 476 pp.

Gans, Carl, ed. *Biology of the Reptilia.* London: Academic Press, 1969--. (5 vols. as of 1976).

Grzimek's Animal Life Encyclopedia: Vol. 5, Fishes II and Amphibians; Vol. 6, Reptiles. New York: Van Nostrand Reinhold, 1972-- 1975.

Peters, James A. *Dictionary of Herpetology.* New York: Hafner, 1964. 392 pp.

Wright, Albert H., and A.A. Wright. *Handbook of Frogs and Toads of the United States and Canada.* 3rd ed. Ithaca, N.Y.: Comstock Pub. Associates, 1949. 640 pp.

Wright, Albert H. *Handbook of Snakes of the United States and Canada.* Ithaca, N.Y.: Com-- stock Pub. Associates, 1957--. 2 vols.

Birds

Campbell, Bruce. *The Dictionary of Birds in Color.* New York: Viking Press, 1974. 352 pp.

Dorst, Jean. *The Life of Birds.* New York: Co-- lumbia University Press, 1974. 2 vols.

**Grzimek's Animal Life Encyclopedia: Vol. 7--9, Birds.* New York: Van Nostrand Reinhold,

1972--1975.

Palmer, Ralph Simon, ed. *Handbook of North American Birds.* New Haven: Yale University Press, 1962--. (3 vols. as of 1976).

Thompson, Arthur L., ed. *A New Dictionary of Birds.* New York: McGraw--Hill, 1964. 928 pp.

Welty, Joel. *The Life of Birds.* 2nd ed. Phila-- delphia: Saunders, 1975. 623 pp.

The World Atlas of Birds. New York: Random House, 1974. 272 pp.

Mammals

Burton, Maurice. *Systematic Dictionary of Mam-- mals of the World.* New York: Crowell, 1962. 307 pp.

**Grzimek's Animal Life Encyclopedia: Vol. 10-- 13, Mammals.* New York: Van Nostrand Reinhold, 1972--1975.

Hall, Eugene Raymond, and K.R. Kelson. *The Mammals of North America.* New York: Ron-- ald, 1959. 2 vols.

Hill, William C.O. *Primates: Comparative Anat-- omy and Taxonomy.* Edinburgh: University Press, 1953--1970. 8 vols.

Napier, John R., and P.H. Napier. *A Handbook of Living Primates.* London: Academic Press, 1967. 456 pp.

Walker, Ernest P. *Mammals of the World.* 3rd ed. Baltimore: Johns Hopkins Press, 1975. 2 vols.

Young, John Z. *The Life of Mammals.* 2nd ed. Oxford, Eng.: Clarendon Press, 1975. 528 pp.

B. Bibliographies

Blackwelder, Richard E. *Guide to the Taxonomic Literature of Vertebrates.* Ames, Iowa: Iowa State University Press, 1972. 259 pp.

Dean, Bashford. *A Bibliography of Fishes.* New York: American Museum of Nautral History, 1916--23; 1968; 1969.

Strong, Reuben Myron. *A Bibliography of Birds.* Chicago: Field Museum of Natural History, 1939–1959.

Walker, Ernest P. *Mammals of the World: Vol. 3, A Classified Bibliography.* Baltimore: Johns Hopkins Press, 1968. 769 pp.

C. Review Serials

International Zoo Yearbook, vol. 1--. London: Zoological Society, 1960--.

E. Catalogs, Checklists, Field Guides

General

Neave, Sheffield A. *Nomenclator Zoologicus: A List of Names of Genera and Subgenera in Zoology.* London: Zoological Society of London, 1939--1940. 4 vols.

------. *Supplements*, vol. 1--, 1950--. (2 vols. as of 1976.

Fish

American Fisheries Society. Committee on Names of Fishes. *A List of Common and Scientific Names of Fishes from the United States and Canada.* Washington, 1970.

Jordan, David. *The Genera of Fishes and A Classification of Fishes.* Stanford, Cal.: Stan-- ford University Press, 1963. 800 pp. (Origin-- ally published 1917–1920).

Lindberg, Georgii U. *Fishes of the World: A Key to Families and a Checklist.* New York: Wiley, 1974. 545 pp.

Reptiles and Amphibians

Cochran, Doris M. and C.J. Goin. *The New Field Book of Reptiles and Amphibians.* New York: Putnam, 1970. 359 pp.

Conant, Roger. *A Field Guide to Reptiles and Amphibians of Eastern and Central North America.* 2nd. ed. Boston: Houghton Mifflin, Mifflin, 1975. 429 pp.

Stebbins, Robert C. *A Field Guide to Western Reptiles and Amphibians.* Boston: Houghton Mifflin, 1966. 279 pp.

Birds

American Ornithologists' Union. *Check--list of North American Birds.* 5th ed. Ithaca, New York, 1957. 691 pp.

Davis, L. Irby. *A Field Guide to the Birds of Mexico and Central America.* Austin: Univer-- sity of Texas Press, 1972. 282 pp.

Edwards, Ernest P. *A Field Guide to the Birds of Mexico.* Sweet Briar, Virginia: E.P. Edwards, 1972. 300 pp.

Peters, James L. *Check--list of Birds of the World.* Cambridge, Mass.: Harvard University Press, 1931--1968. 15 vols.

For a fuller list of keys and other taxonomic works see Roger C. Smith, *Guide to the Litera-- ture of the Life Sciences.* Minneapolis: Burgess, 1972.

G. Indexes

Wildlife Review: An Abstracting Service for Wildlife Management, vol. 1--. Fort Collins,

Colo.: Colorado State University, 1935--.

16. Parasitology

A. Encyclopedias, Treatises, Texts, Dictionaries, Handbooks

Noble, Elmer R. and G.A. Noble. *Parasitology: The Biology of Animal Parasites.* 3rd ed. Philadelphia: Lea & Febiger, 1971. 617 pp.

Olsen, Oliver. *Animal Parasites: Their Life Cycles and Ecology.* 3rd ed. Baltimore: University Park Press, 1974. 562 pp.

C. Review Serials

Advances in Parasitology, vol. 1--. London: Academic Press, 1963--.

G. Indexes

Helminthological Abstracts: A Quarterly Review of Literature on Helminths and Their Vectors. vol. 1--. St. Albans, England: Imperial Bureau of Agriculture Parasitology, 1932--.

17. Marine Biology (See also plant and animal headings for groups of marine species.)

A. Encyclopedias, Treatises, Texts, Dictionaries, Handbooks

Fairbridge, Rhodes Whitmore. *The Encyclopedia of Oceanography.* New York: Reinhold, 1966. 1021 pp.

Firth, Frank E. *The Encyclopedia of Marine Resources.* New York: Van Nostrand Reinhold, 1969. 740 pp.

Food and Agriculturel Organization of the United Nations. *Atlas of the Living Resources of the Sea.* 3rd ed. Rome, 1972. 1 v. (various paging)

Kinne, Otto. *Marine Ecology: a Comprehensive, Integrated Treatise on Life in Oceans and Coastal Waters.* London: Wiley, 1970--. (9 vol. as of 1976.)

C. Review Serials

Advances in Marine Biology, vol. 1--. London: Academic Press, 1963--.

Oceanography and Marine Biology, vol. 1--. London: Allen and Unwin, Ltd., 1963--.

18. Agriculture

A. Encyclopedias, Treatises, Texts, Dictionaries, Handbooks

The Merck Veterinary Manual; a Handbook of Diagnosis and Therapy for the Veterinarian, 4th ed. Rahway, N.J.: Merck, 1973. 1618 pp.

C. Review Serials

Advances in Agronomy, vol. 1--. New York: Academic Press, 1949--.

H. Guides to the Literature

Blanchard, Joy Richard, and H. Ostvold. *Literature of Agricultural Research.* Berkeley: University of California Press, 1958. 231 pp.

19. Economic Botany

A. Encyclopedias, Treatises, Texts, Dictionaries, Handbooks

Uphof, Johannes T. *Dictionary of Economic Plants.* 2nd ed. New York: Hafner, 1968. 591 pp.

Usher, George. *A Dictionary of Plants Used by Man.* New York: Hafner, 1974. 619 pp.

20. Environmental Science; Pollution

A. Encyclopedias, Treatises, Texts, Dictionaries, Handbooks

Fairbridge, Rhodes, ed. *The Encyclopedia of Geochemistry and Environmental Sciences.* New York: Van Nostrand Reinhold, 1972. 1321 pp.

McGraw--Hill Encyclopedia of Environmental Science. New York: McGraw--Hill, 1974. 754 pp.

Stern, Arthur Cecil. *Air Pollution.* 3rd ed. New York: Academic Press, 1976. 4 vols.

U.S. National Science Foundation, National Science Board. *Patterns and Perspectives in Environmental Science.* Washington, D.C.: U.S. Government Printing Office, 1973. 426 pp.

Sarnoff, Paul. *The New York Times Encyclopedic Dictionary of the Environment.* New York: Quadrangle Books, 1971. 352 pp.

Todd, David, ed. *The Water Encyclopedia; a Compendium of Useful Information on Water Resources.* Port Washington, New York: Water Information Center, 1970. 559 pp.

Water and Water Pollution Handbook. Ed. Leonard L. Ciaccio. New York: Dekker, 1971--73. 4 vols.

B. Bibliographies

National Foundation for Environmental Control. *NFEC Directory of Environmental Information Sources.* 2nd ed. Boston, 1972. 457 pp.

Onyx Group, Inc. *Environment U.S.A.: a Guide to Agencies, People, and Resources.* New York: Bowker, 1974. 451 pp.

Harrah, Barbara, and D. Harrah. *Alternate Sources of Energy: a Bibliography of Solar, Geothermal, Wind, and Tidal Energy, and Environmental Architecture.* Metuchen, N.J.: Scarecrow, 1975. 201 pp.

Thomas, William A., W.H. Wilcox, and G. Gold--stein. *Biological Indicators of Environmental Quality: A Bibliography of Abstracts.* Ann Arbor, Mich.: Ann Arbor Science Publishers, 1973. 254 pp.

Tompkins, Dorothy. *Strip Mining for Coal.* Insti-tute of Governmental Studies, University of California, 1973. 86 pp.

U.S. Division of Air Pollution. *Air Pollution Publications: A Selected Bibliography, 1955/62--68.* Washington, D.C., Public Health Service, Div. of Air Pollution, 1963–1969. 2 vol.

Winton, Harry N.M. *Man and the Environment: A Bibliography of Selected Publications of the United Nations System, 1946--1971.* New York: Unipub, 1972. 305 pp.

C. Review Serials

Advances in Environmental Science and Tech-nology, vol. 1--. New York: Wiley, 1969.

CRC. Critical Reviews in Environmental Con-trol, vol. 1–. Cleveland, Ohio: CRC Press, 1970.

D. Data Compendia

CRC. Handbook of Environmental Control. Cleveland, Ohio: CRC Press, 1972--1975. 5 vol.

F. Methods

American Society for Testing and Materials. Committee D--19 on Water. *Manual on Water.* 3rd ed. Philadelphia, Amer. Soc. for Testing and Materials, 1969. 360 pp.

G. Indexes

Air Pollution Abstracts, vol. 1--. Research Triangle Park, N.C.: Air Pollution Technical Information Center, 1969--.

Environment Abstracts, vol. 1–. New York: En-vironment Information Center, Inc., 1971–.

Environment Index: A Guide to the Key En-vironmental Literature of the Year. New York: Environment Information Center, Inc. 1972--.

Pollution Abstracts, vol. 1--. Louisville, Ky.: Data Courier, 1970–.

Selected References on Environmental Quality as it Relates to Health, vol. 1--. Bethesda, Md.: National Library of Medicine, 1971–.

21. Medicine

A. Encyclopedias, Treatises, Texts, Dictionaries, Handbooks

Black's Medical Dictionary. 31st ed. London: A. & C. Black, 1976. 950 pp.

Cecil, Russell La Fayette. *Textbook of Medicine.* 14th ed. Philadelphia: Saunders, 1975. 2 vols.

The United States Dispensatory. 27th ed. Phila-delphia: Lippincott, 1973. 1292 pp.

Handbook of Clinical Laboratory Data. Cleveland, Ohio: Chemical Rubber Co., 1968. 710 pp.

The Merck Index: An Encyclopedia of Chemicals and Drugs. 9th ed. Rahway, N.J.: Merck, 1976. 1 vol.

The Merck Manual of Diagnosis and Therapy. 12th ed. Rahway, N.J.: Merck, 1972. 1964 pp.

Physicians' Desk Reference to Pharmaceutical Specialties and Biologicals. 1st ed. Oradell, N.J.: Medical Economics, 1947--. (30th ed., 1976).

Stedman, Thomas Lathrop. *Stedman's Medical Dictionary.* 23rd ed. Baltimore: Williams & Wilkins Co., 1976. 1678 pp.

B. Bibliographies

U.S. National Library of Medicine. *Bibliography of Medical Reviews; Monthly Bibliography of Medical Reviews,* vol. 1--. Washington, D.C., 1955–1971.

C. Review Serials

Advances in Drug Research, vol. 1–. New York: Academic Press, 1964–.

Advances in Pharmacology and Chemotherapy, vol. 1–. New York: Academic Press, 1962–.

Annual Reports in Medicinal Chemistry, vol. 1--. New York: Academic Press, 1965–.

Annual Review of Medicine, vol. 1--. Palo Alto, Cal.: Annual Reviews, Inc., 1950–.

Annual Review of Pharmacology and Toxicology, vol. 1--. Palo Alto, Cal.: Annual Reviews, 1961--.

Nutrition Reviews, vol. 1–. New York: Nutrition Foundation, Inc., 1942--.

Perspectives in Biology and Medicine, vol. 1–.

Chicago: University of Chicago Press, 1957--.
Pharmacological Reviews, vol. 1--. Baltimore: Williams & Wilkins Co., 1949--.
Progress in Medical Genetics, vol. 1--10. New York: Drune & Stratton, 1961–74.
Progress in Medical Genetics New Series, vol. 1–. New York: Saunders, 1976--.
Progress in Medical Virology, vol. 1--. New York: S. Karger, 1958--.
Progress in Medicinal Chemistry, vol. 1--. London: Butterworths, 1961--.

D. Data Compendia

International Commission on Radiological Pro-- tection. Committee II on Permissible Dose for Internal Radiation Task Group. *Report of the Task Group on Reference Man.* Oxford, Eng.: Pergamon Press, 1975. 480 pp.

F. Methods

Methodology of Analytical Toxicology. Cleveland, Ohio: Chemical Rubber Co., 1975. 478 pp.
Manual of Analytical Toxicology. Cleveland, Ohio: Chemical Rubber Co., 1971. 893 pp.
Manual of Clinical Laboratory Procedures. 2nd ed. Cleveland, Ohio: Chemical Rubber Co. 354 pp.
Abridged Index Medicus, vol. 1–. Bethesda, Md.: National Library of Medicine, 1970–.
Index Medicus, vol. 1–. Bethesda, Md.: National Library of Medicine, 1960--.
Psychopharmacology Abstracts, vol. 1--. Chevy Chase, Md.: National Clearinghouse for Mental Health Information, 1961--.

H. Guides to the Literature

Blake, John B. and C. Roos. *Medical Reference Works, 1679--1966: A Selected Bibliography and Supplements 1 and 2.* Chicago: Medical Library Association, 1967--73. 3 vols.
Brunn, Alice L. *How to Find Out in Pharmacy: A Guide to Sources of Pharmaceutical Infor-- mation.* New York: Pergamon Press, 1969. 130 pp.

APPENDIX V

Using *Chemical Abstracts*

Why another indexing/abstracting service?

The sample topic used in this book did not have a "chemical angle" to it and therefore it was not necessary, or possible, to illustrate the use of *C.A.* to find information on the topic. This does not mean that *Chemical Abstracts* is less important than *Biological Abstracts.* It is not appropriate to think of any one index as more important to a biologist than any other. Instead users should reflect on what literature it is that you want and then ask yourself, "Which index covers the literature I want?" Unfortunately just because you know the question does not mean you know the answer. It takes many years' experience to learn which index is best. A general rule of thumb is to use both *Chemical Abstracts* and *Biological Abstracts* if your topic has any chemical, biochemical or physiological aspect to it; but the wise will consult with a reference librarian about which indexing/abstracting service to use.

What is it and how do I use it?

Since you are already familiar with *Biological Abstracts,* you are already aware of *C.A.*'s basic organization. However, there are some differences between *B.A.* and *C.A.*, and the strategies for use are significantly different. What follows is not a thorough introduction to *C.A.* For a more comprehensive introduction you should consult one of the guides to the literature (Appendix VII) or the "Introduction" at the beginning of each index which *C.A.* publishes. What follows here is an outline of the organization of *C.A.* and some suggestions about how biologists might use it.

Weekly Issue

C.A. is weekly. Like *B.A.* the abstracts are grouped by subject. However it takes two weekly issues to cover the eighty subject categories. FIGURE C shows these subject categories as they are listed in the Tables of Contents. Each weekly issue contains, in addition to the abstracts, the following indexes: "Keyword," "Numerical Patent," "Patent Concordance," and "Author." As a biologist you will primarily use the author index to locate recent articles by the persons who are writing in your field of study, or the "Keyword Index" for subject areas. While the "Keyword Index" of *C.A.* does not look like that of *B.A.* (see FIGURE D for *C.A.*'s "Keyword Index"), it should be used in the same way. That is you must use the vocabulary of the workers in the field; search all synonyms, related terms, and all related aspects. Once you have located articles that appear relevant check the abstract number in the first part of the issue.

Six Months and Cumulative Issues

Twice a year *C.A.* produces a series of indexes which cover the abstracts of the previous six months. Then these indexes are cumulated every ten years and more recently every five years. Because the indexes are somewhat different than those in the weekly issue and because there are a greater number of indexes it is best not to compare the six month and cumulation indexes with the weekly ones.

In Table 1 are listed the available cumulative indexes. With this number of indexes you can search *C.A.* in a number of different ways. If you are unsure about using *C.A.* indexes ask your reference librarian how to search it for the information you want. As a biologist, the index that you will most likely use is the "Subject Index." This index has been carefully constructed in order to bring all the material on a specific subject together. Therefore you can expect to find all material on a specific compound at the same spot (i.e. articles will not be scattered under listings which depend on what name the author used). This does not, however, solve the problem of hierarchical relationships, and therefore it is important to check more general headings if the specific one does not give appropriate information.

Before 1967 the "Subject Index" contained all the Cross References and Notes which the indexers had built into the system. Beginning

	1st Coll. 1907--1926	2nd Coll. 1917–1926	3rd Coll. 1927–1936	4th Coll. 1937–1946	5th Coll. 1947–1956	6th Coll. 1957–1961	7th Coll. 1962–1966	8th Coll. 1967–1971	9th Coll. 1972–1976	Current
AUTHOR	Yes	Yes	Yes	Yes	Yes	Yes	Yes	Yes	Yes	Yes
FORMULA	No	No [Yes, 1920--1946 ----------------------]			Yes	Yes	Yes	Yes	Yes	Yes
INDEX OF RING SYSTEMS	No	No	No	[Yes, 1940--1963]				No	Yes	No
PATENT										
NUMERICAL	No	No	No	Yes	Yes	Yes	Yes	Yes	Yes	Yes
CONCORDANCE	No	No	No	No	No	No	Yes	Yes	Yes	Yes
REGISTRY NUMBER	No	No	No	No	No	No	No [Yes, 1965----1971]		Yes	Yes
SUBJECT	Yes	Yes	Yes	Yes	Yes	Yes	Yes	Yes	No	No
GENERAL SUBJECT	No	No	No	No	No	No	No	No	Yes	Yes
CHEM SUBSTANCES	No	No	No	No	No	No	No	No	Yes	Yes
INDEX GUIDE	No	No	No	No	No	No	No	Yes	Yes	Yes

TABLE 1. History of indexes to Chemical Abstracts.

CHEMICAL ABSTRACTS

KEY TO THE WORLD'S CHEMICAL LITERATURE

A Publication of the CHEMICAL ABSTRACTS SERVICE

Published weekly by the AMERICAN CHEMICAL SOCIETY, *2540 Olentangy River Rd., Columbus, Ohio 43202*

Editorial Office: Chemical Abstracts Service, P.O. Box 3012, Columbus, Ohio 43210 Telephone (614) 421-6940 Teletype 810 482-1608

Chemical Abstracts, a publication of the Chemical Abstracts Servic., is the member journal for chemistry in the English language on the Abstracting Board of the International Council of Scientific Unions. CAS is also a member of the National Federation of Ab= stracting and Indexing Services in America.

It is the careful endeavor of *Chemical Abstracts* to publish adequate and accurate abstracts of all scientific and technical papers containing new information of chemical and chemical engineering interest and to report new chemical information revealed in the patent literature, but the American Chemical Society is not responsible for omissions or for such mistakes as may be made in abstracts and index entries.

All questions regarding editorial content should be addressed to the Editorial Office, Chemical Abstracts Service, P. O. Box 3012, Columbus, Ohio 43210.

Second-class postage paid at Columbus, Ohio and additional office.
Library of Congress Catalog Card Number: 9-4698

SUBSCRIPTION INFORMATION

For residents of the United States and countries other than the United Kingdom of Great Britain and Northern Ireland, the Republic of Ireland, and the Federal Republic of Germany, the Annual Subscription Rates for 1977 (Volumes 86 and 87) are:

Abstract Issues with Volume Indexes $3500.00; Abstract Issues only or Volume Indexes only $3350.00.

Postage per year: Abstract Issues with Volume Indexes - U.S. none, PUAS $57.00, others $68.00; Abstract Issues without Volume In= dexes--U.S. none, PUAS $37.00, others $44.00; Volume Indexes only--U.S. none, PUAS $20.00, others $24.00.

A one year's subscription may begin in January or July.

Prices of single issues (except indexes) from these volumes: $60.00. Postage for single issues: U.S. none; PUAS and other foreign $3.00.

Degree granting educational institutions may be eligible for a $500.00 grant toward a current subscription to *Chemical Abstracts*. For further information, contact the appropriate organization listed below.

Claims for missing printed issues from residents of the United States, Canada, and Mexico must be submitted within 90 days; from all other countries, within one year. No claims allowed because of failure to report a change of address or because copy is "missing from files".

Report a subscription address change immediately. Include the following information: (1) name of printed publication(s); (2) Customer Account Number (first 8 digits on address label); (3) old mailing address for each publication (include a recent mailing label when possible); (4) new mailing address with ZIP Code or equivalent for each publication.

Forty of the 80 *CA* Sections are available in computer-readable, magnetic tape form as six individual, topic-oriented information services: *Chemical Biological Activities, Ecology and Environment, Energy, Food and Agricultural Chemistry, Materials*, and *Polymer Science and Technology. CA* Heading and Keyword data are available in computer readable, magnetic tape form on the information service *CA Condensates*. Contact the appropriate organization listed below for further information.

Depending on country of residence, questions regarding business transactions, subscriptions, and services of Chemical Abstracts Service, including back issues, should be directed to the following organizations:

1. Residents of the **United Kingdom of Great Britain and Northern Ireland** and the **Republic of Ireland** should contact the United Kingdom Chemical Information Service, The University, Nottingham NG7 2RD, England.
2. Residents of the **Federal Republic of Germany** should contact Verlag Chemie, Postfach 1260/1280, D-6940 Weinheim, Federal Republic of Germany.
3. Residents of the **United States** and **other countries** except the United Kingdom of Great Britain and Northern Ireland and the Republic of Ireland and the Federal Republic of Germany should contact the following departments at Chemical Abstracts Service, P. O. Box 3012, Columbus, Ohio 43210: (Telephone (614) 421 6940; Teletype 810 482 1608).
   - a. ***Subscription Service Department*** **for business transactions including subscription information, remittances, change of address, and claims for missing issues.**
   - b. ***Marketing Department*** **for information regarding CAS services.**

Volume 87 • Number 1 • July 4, 1977

CONTENTS

CODEN: CHABA8 87(1) 1-521(1977)
ISSN: 0009-2258

ABSTRACT SECTIONS

Biochemistry Sections

Organic Chemistry Sections

ISSUE INDEXES

Indexes to this issue are found at the back of this issue in the order listed below:

Sections 35-80 covering Macromolecular Chemistry, Applied Chemistry and Chemical Engineering, and Physical and Analytical Chemistry appear in alternate issues of *Chemical Abstracts*.

CHEMICAL ABSTRACTS

KEY TO THE WORLD'S CHEMICAL LITERATURE

A Publication of the CHEMICAL ABSTRACTS SERVICE

Published weekly by the AMERICAN CHEMICAL SOCIETY, *2540 Olentangy River Rd., Columbus, Ohio 43202*

Editorial Office: Chemical Abstracts Service, P.O. Box 3012, Columbus, Ohio 43210 Telephone (614) 421-6940 Teletype 810 482-1608

Chemical Abstracts, a publication of the Chemical Abstracts Service, is the member journal for chemistry in the English language on the Abstracting Board of the International Council of Scientific Unions. CAS is also a member of the National Federation of Ab= stracting and Indexing Services in America.

It is the careful endeavor of *Chemical Abstracts* to publish adequate and accurate abstracts of all scientific and technical papers containing new information of chemical and chemical engineering interest and to report new chemical information revealed in the patent literature, but the American Chemical Society is not responsible for omissions or for such mistakes as may be made in abstracts and index entries.

All questions regarding editorial content should be addressed to the Editorial Office, Chemical Abstracts Service, P. O. Box 3012, Columbus, Ohio 43210.

Second-class postage paid at Columbus, Ohio and additional office.
Library of Congress Catalog Card Number: 9 4698

SUBSCRIPTION INFORMATION

For residents of the United States and countries other than the United Kingdom of Great Britain and Northern Ireland, the Republic of Ireland, and the Federal Republic of Germany, the Annual Subscription Rates for 1977 (Volumes 86 and 87) are:

Abstract Issues with Volume Indexes $3500.00; Abstract Issues only or Volume Indexes only $3350.00.

Postage per year: Abstract Issues with Volume Indexes U.S. none, PUAS $57.00, others $68.00; Abstract Issues without Volume In= dexes--U.S. none, PUAS $37.00, others $44.00; Volume Indexes only--U.S. none, PUAS $20.00, other $24.00.

A one-year's subscription may begin in January or July.

Prices of single issues (except indexes) from these volumes: $60.00. Postage for single issues: U.S. none; PUAS and other foreign $3.00.

Degree-granting educational institutions may be eligible for a $500.00 grant toward a current subscription to *Chemical Abstracts*. For further information, contact the appropriate organization listed below.

Claims for missing printed issues from residents of the United States, Canada, and Mexico must be submitted within 90 days; from all other countries, within one year. No claims allowed because of failure to report a change of address or because copy is "missing from files".

Report a subscription address change immediately. Include the following information: (1) name of printed publication(s); (2) Customer Account Number (first 8 digits on address label); (3) old mailing address for each publication (include a recent mailing label when possible); (4) new mailing address with ZIP Code or equivalent for each publication.

Forty of the 80 *CA* Sections are available in computer-readable, magnetic tape form as six individual, topic-oriented information services: *Chemical-Biological Activities, Ecology and Environment, Energy, Food and Agricultural Chemistry, Materials*, and *Polymer Science and Technology. CA* Heading and Keyword data are available in computer-readable, magnetic tape form on the information service *CA Condensates*. Contact the appropriate organization listed below for further information.

Depending on country of residence, questions regarding business transactions, subscriptions, and services of Chemical Abstracts Service, including back issues, should be directed to the following organizations:

1. Residents of the **United Kingdom of Great Britain and Northern Ireland** and the **Republic of Ireland** should contact the United Kingdom Chemical Information Service, The University, Nottingham NG7 2RD, England.
2. Residents of the **Federal Republic of Germany** should contact Verlag Chemie, Postfach 1260/1280, D-6940 Weinheim, Federal Republic of Germany.
3. **Residents of the United States and other countries except the United Kingdom of Great Britain and Northern Ireland and the Republic of Ireland and the Federal Republic of Germany should contact the following departments at Chemical Abstracts Service, P. O. Box 3012, Columbus, Ohio 43210: (Telephone (614) 421-6940; Teletype 810 482-1608).**
   - a. ***Subscription Service Department*** **for business transactions including subscription information, remittances, change of address, and claims for missing issues.**
   - b. ***Marketing Department*** **for information regarding CAS services.**

Volume 87 • Number 2 • July 11, 1977

CODEN: CHABA8 87(2) 1-672(1977)
ISSN: 0009-2258

ABSTRACT SECTIONS

Macromolecular Chemistry Sections

Applied Chemistry and Chemical Engineering Sections

Physical and Analytical Chemistry Sections

ISSUE INDEXES

Indexes to this issue are found at the back of this issue in the order listed below:

CA Abstracted Publications: Additions and Changes

Sections 1-34 covering Biochemistry and Organic Chemistry appear in alternate issues of *Chemical Abstracts*.

FIGURE C. Chemical Abstracts, Tables of Contents.

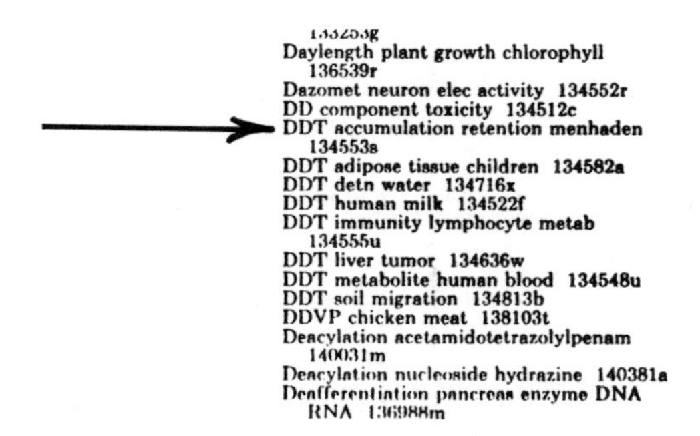

Daylength plant growth chlorophyll 136539r
Dazomet neuron elec activity 134552r
DD component toxicity 134512c
DDT accumulation retention menhaden 134553s
DDT adipose tissue children 134582a
DDT detn water 134716x
DDT human milk 134522f
DDT immunity lymphocyte metab 134555u
DDT liver tumor 134636w
DDT metabolite human blood 134548u
DDT soil migration 134813b
DDVP chicken meat 138103t
Deacylation acetamidotetrazolylpenam 140031m
Deacylation nucleoside hydrazine 140381a
Dedifferentiation pancreas enzyme DNA RNA 136988m

FIGURE D. Chemical Abstracts, weekly "Keyword Index", vol. 86, no. 19, 1977.

in 1967 these finding aids were relegated to the separate "Index Guide." The astute reader might conclude that this "Index Guide" functions, in relation to the index, in much the same way as the *Library of Congress Subject Headings* does in relation to the card catalog. And to a large extent that is true. The most significant difference is that *not* all headings are in the "Index Guide"; only those headings which have cross references or notes are included.

The two major uses of the subject indexes will probably be for material on a specific compound or a particular organism. In either case begin with the most recent Collective Index. Search for the compound in the "Chemical Substances Index," an organism in the "General Subjects Index." The compound should be looked up under the established name (see the Introduction to *C.A.*'s index for details), in inverted form (e.g. 1, 2 dichlorobenzene under Benzene, 1, 2 dichloro). Organisms should be searched under their scientific name (e.g. White-tailed deer under *Odocoileus virginianus*).

If you do not find your compound or organism in the subject indexes then you should consult the "Index Guide," for a cross reference to the appropriate heading. For example, if you are working on "3-butenyl methy ketone," a search of the "Chemical Substances Index" under ketone will reveal nothing. However, a check of the "8th Collective Index Guide" will help you out. FIGURE E shows you the entry for Ketone, 3-butenyl methyl. The "Index Guide" should also be used when you have a colloquial name, e.g. DDT, or a proprietary name, e.g. Teflon, (FIGURE E). If the "Index Guide" does not give you any information then you need to take some additional steps.

1. If you have the formula for your com-

DDS (pharmaceutical)
See *Aniline, 4,4' - sulfonyldi* - [*80 - 08 - 0*]
DD Soil Fumigant
See *Propene, 1,3 - dichloro* - [*542 - 75 - 6*], mixt. with 1,2-dichloropropane
DDT
See *Ethane, 1,1,1 - trichloro - 2,2 - bis(p - chlorophenyl)* - [*50 - 29 - 3*]
o,p'-DDT
See *Ethane, 1,1,1 - trichloro - 2 -(o - chlorophenyl) - 2 -(p - chlorophenyl)* - [*789 - 02 - 6*]

Ketomycin
See *3 - Cyclohexene - 1 - glyoxylic acid, (R)* - [*23364 - 22 - 9*]
Ketone
methyl phenylazo—See *Diimide, acetylphenyl* - [*13443 - 97 - 5*]
—, **1-acetylindolyl methyl**
See *Indole, C,1 - diacetyl* -
—, **anthryl ethyl**
See *1 - Propanone, 1 -(anthryl)* -
—, **benzodioxanyl methyl**
O -(2 -cyanoethyl)oxime—see *Propionitrile, 3 -[[[1 -(benzodioxanyl)ethylidene]amino]oxy]* -
—, **benzofuranyl ethyl**
See *1 - Propanone, 1 -(benzofuranyl)* -
—, **benzofuranyl styryl**
See *2 - Propen - 1 - one, 1 -(benzofuranyl) - 3 - phenyl* -
—, **benzyl ethyl**
See *2 - Butanone, 1 - phenyl* - [*1007 - 32 - 5*]
—, **benzyl methyl**
See *2 - Propanone, 1 - phenyl* - [*103 - 79 - 7*]
—, **benzyl naphthyl**
See *Acetonaphthone, 2 - phenyl* -
—, **benzyl phenyl**
See *Acetophenone, 2 - phenyl* - [*451 - 40 - 1*]
—, **biphenylyl ethyl**
See *Propiophenone, ar - phenyl* -
—, **4-biphenylyl *α*-hydroxy-*p*-phenylbenzyl**
See *Benzoin, 4,4' - diphenyl* - [*5623 - 25 - 6*]
—, **biphenylyl methyl**
See *Acetophenone, ar - phenyl* -
—, **biphenylyl phenyl**
See *Benzophenone, phenyl* -
—, **4-biphenylyl *p*-phenylbenzyl**
See *Acetophenone, 2 -(4 - biphenylyl) - 4' - phenyl* - [*1694 - 24 - 2*]
—, **bis(2-benzothiazolyl)**
See *2 - Benzothiazolyl ketone* [*4464 - 60 - 2*]
—, **bis(benzotriazolyl)**
See *Benzotriazolyl ketone*
—, **bis(methoxyphenyl)**
See *Benzophenone, ar,ar' - dimethoxy* -
—, **9-bromo-6,10,10-trimethylspiro[4.5]dec-6-en-2-yl methyl**
(-) - (*spirolaurenone*)
—, **3-butenyl methyl**
See *5 - Hexen - 2 - one* [*109 - 49 - 9*]
—, **butenyl phenyl**
See *Pentenophenone*
—, **butyl furyl**
See *1 - Pentanone, 1 -(furyl)* -
—, **butyl methyl**
See *2 - Hexanone* [*591 - 78 - 6*]
—, ***tert*-butyl methyl**
See *2 - Butanone, 3,3 - dimethyl* - [*75 - 97 - 8*]
—, **butyl naphthyl**

Tedion
See *Sulfone, p - chlorophenyl 2,4,5 - trichlorophenyl* [*116 - 29 - 0*]
Tedlar
See *Ethylene, fluoro* -, *polymers*, polymers [*24981 - 14 - 4*], Tedlar
Teeth
See also
Dental materials
Dentifrices
artificial—see *Dentures*
fluorosis of—see *Teeth*, mottled, enamel of
Tefamin
See *Theophylline*, compd. with ethylenediamine (2:1) [*317 - 34 - 0*]
Teflon
See *Ethylene, tetrafluoro* -, *polymers*, polymers [*9002 - 84 - 0*], Teflon
Teflon FEP
See *Propene, hexafluoro* -, polymer with tetrafluoroethylene [*25067 - 11 - 2*], Teflon FEP
Teflon FEP 100A
See *Propene, hexafluoro* -, polymer with tetrafluoroethylene [*25067 - 11 - 2*], Teflon FEP 100A

FIGURE E. Chemical Abstracts, "8th Collective Index Guide, 1967 - 71", pp. 557, 1182, 1915.

pound then use the "Formula Index." For example, the compound chlorazone is not listed in the "Index Guide" or "Chemical Substances Index" by that name. If you know the formula is $C_7H_7ClNNaO_2S$, you can look for it in the "Formula Index" (FIGURE F.)

C_7H_7ClMgO
Magnesium, ch! ...oxyphenyl)-, 56:P 5031; 62:11839g
$C_7H_7ClNNaO_2S$
(See also *Sodium, (N-chloro-p-toluenesulfonamido)-.*)
Sodium, (chloro-*p*-toluenesulfonamido)-, 60:8375a
—, (*m*-chloro-*p*-toluenesulfonamido)-, 61:13647c
$C_7H_7ClNO_3PS$
Phosphonochloridothioic acid, methyl-, *O*-*p*-nitrophenyl ester, 58:546g

FIGURE F. Chemical Abstracts, "7th Collective Formula Index, 1962 - 66", p. 608.

2. If you do not know the formula then you need to consult basic reference sources in chemistry (Appendix IV). These hopefully will give you other names, one of which should show up in the "Index Guide" or "Chemical Substances Index."
3. If neither of the steps is useful ask a reference librarian for help.

An example of the "Chemical Substances Index" is shown in FIGURE G.
In FIGURE G, note the following:

1. The number [50–29–3]. This is a "Registry Number" which is unique to each compound and which does not change although the name by which the compound is indexed may. For example, DDT is listed under a different name in the 1972–76 index, but the number remains the same. This number becomes a convenient method of checking whether you have the correct entry. This "Registry Number" is becoming increasingly important in the chemical literature to the point where some authors now use the number in their publications.
2. Since there are a large number of articles on DDT the articles have been grouped into categories: Analysis, Biological Studies, Occurence, Preparation, Properties, Reactions, Uses and Miscellaneous, and Compounds.
3. Each line under the heading represents a different abstract. These lines are

Ethane, 1,1,1-trichloro-2,2-bis(*p*-chlorophenyl)- [*50 - 29 - 3*], analysis
69: 51124a
adsorption of, by carbon, **74:** 86706x
biodegradability of, **75:** 4463z
in blood after single intensive exposure, **71:** 53279v
chlorine detn. in, **69:** 2062v

Ethane, 1,1,1-trichloro-2,2-bis(*p*-chlorophenyl)- [*50 - 29 - 3*], biological studies
69: 18097w; **71:** 100562u
absorption and distribution and metabolism of, in chickens, **69:** 95376p
absorption and elimination of, by *Salmo gairdnerii*, **74:** 75320r

accumulation of
in adipose tissue and organs, **73:** 76074e
in aquatic environment, **74:** R 52404g — Review
by *Daphnia magna*, **75:** 75302g
in environment, **75:** 97550p
in fauna after field application, **71:** 69643v
in human organism, **72:** 65783p
in laying hens after feeding, **72:** 131418b
in lentic environment, adsorption in, **75:** 75284c
in natural wastes, **70:** 67146r
by nymph of *Tetragoneuria*, **75:** 87453v
in ocean waters, **75:** R 75232j
in tapeworms from mallard and lesser scaup ducks, **73:** 65344s

biol. synergists for, **74:** R 22068h
bird mortality from, in Netherlands, **69:** 75896e
bird poisoning by, in orchards, **69:** 43026e
in birds and fish and invertebrates of Pacific, **68:** 38512k
in birds of South Dakota, **71:** 12064c
bis(chlorophenyl)acetaldehyde in metabolism of, **71:** 112564j
for *Biston suppressaria* control on tea, **67:** 63256s
in black and mallard ducks, **71:** 111842m
Blastophagus piniperda control by, **75:** 47795b
in blood

Dysdercus fasciatus response to sublethal doses of, **75:** 34450k
Dysdercus kœnigii in response to, **66:** 18272n
in eagles, **72:** 65875v
in earthworms, **68:** 113575p
of earthworms and *Turdus migratorius*, **73:** 130096d
ecology of salmon streams in Alaska after spraying with, **66:** 85058k
in ecosystem of Antarctic, **70:** 67150n
in ecosystem of St. Lawrence River, **73:** 2962w
Edwardsiana control by, on apples, **72:** 110230u

eggshell thickness of eggs of falcons and hawks in relation to, **75:** 75329w
in eggs of birds of prey, **70:** 86543j
of eggs of *Haliaeetus leucocephalus*, **75:** 75310h
of eggs of Norway, **75:** 117231r

Ethane, 1,1,1-trichloro-2,2-bis(*p*-chlorophenyl)- [*50 - 29 - 3*], occurrence
in water, of Mississippi and Missouri Rivers, **72:** 6161m
Ethane, 1,1,1-trichloro-2,2-bis(*p*-chlorophenyl)- [*50 - 29 - 3*], preparation
manuf. of, **69:** P 43601p — Patent
of powd., **66:** P 45783m
purification of, **72:** P 43093m
volatilization of, soil colloid effect on, **68:** 77192a
waste water from, *p*-chlorobenzenesulfonic acid adsorption from, **68:** 107738c
Ethane, 1,1,1-trichloro-2,2-bis(*p*-chlorophenyl)- [*50 - 29 - 3*], properties
adsorption by soil, org. matter effect on, **70:** 10586e

Ethane, 1,1,1-trichloro-2,2-bis(*p*-chlorophenyl)- [*50 - 29 - 3*], reactions
chlorination of, by phosphorus chloride, **72:** 12239j
dechlorination of
in aq. media under pressure, **72:** 120407w
by gamma-irradiation, **75:** 117814h
by ionizing radiation, **75:** 109106p

Ethane, 1,1,1-trichloro-2,2-bis(*p*-chlorophenyl)- [*50 - 29 - 3*], uses and miscellaneous
coatings contg., **67:** 65467k
detection of uv degradtion products of, **68:** 103905g
dimethyl dichlorovinyl phosphate detn. in presence of, **75:** 109048w

Ethane, 1,1,1-trichloro-2,2-bis(*p*-chlorophenyl)-, compounds
compd.

FIGURE G. Chemical Abstracts, "8th Collective Subject Index, 1967 - 71", pp. 11609 - 11611. (Not shown: B, stands for Book)

(3) **Evolution**
of albumins, generation time of animal in relation to, 150105s
amino acid and nucleotide sequence in relation to, R 30392m
of angiosperms
45647x
cytochrome c in detn. of, 162102b
in angiotensin-renin active components, 44554c
of bats, cytochrome c in relation to, 136489p
biochem., earth history in relation to, R 84553h
blood properties of black and ringtailed lemurs in relation to, 150110q
chem.
R 15634y
of peptides, aminoacetonitrile in relation to, 44623z
phosphorus in prebiotic, R 84551f
chemical, R 161062b

(1) **Halfordia kendack**
alkaloids of, synthesis of, 34738b
Haliaeetus leucocephalus
eggs of, chlorine-contg. insecticides and mercury of, 136004b
Halibut
exts., trimethylamine and volatile nitrogen of cooked, *Pseudomonas fragi* in relation to, 138438v

(2) **Kidney**
acid-base equil. of urine regulation by, 137745f
adaptation to hypocapnia, 59466z
aflatoxin and sterigmatocystin analogs effect on cultures of, 110310y
aldosterone receptors of, 109773v
alkalosis maintenance by, R 109815k
alloplastic prosthetic materials effect on, 1490m
allotransplantation of, azathioprine in survival of, circulation in relation to, 96884e
allotransplant of, immunosuppressants effect on survival of, 506j
aminohippurate clearance by, in newborn, 59543x
angiotensin analogs effect on, 29317j
angiotensin and and mannitol effect on, toxicity in relation to, 70387v
angiotensinogen of blood plasma in relation to, 3305d
antibodies to, in immunoglobulins to lymphocytes, 86540a
antibody-Gross leukemia virus complexes and complement in glomeruli of, in leukemia, 124650w

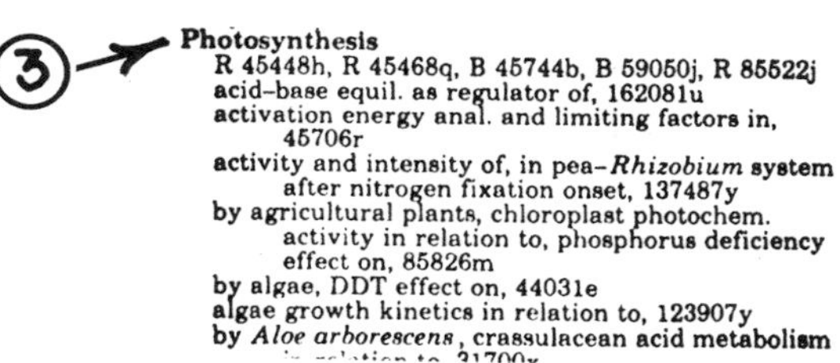
(3) **Photosynthesis**
R 45448h, R 45468q, B 45744b, B 59050j, R 85522j
acid-base equil. as regulator of, 162081u
activation energy anal. and limiting factors in, 45706r
activity and intensity of, in pea-*Rhizobium* system after nitrogen fixation onset, 137487y
by agricultural plants, chloroplast photochem. activity in relation to, phosphorus deficiency effect on, 85826m
by algae, DDT effect on, 44031e
algae growth kinetics in relation to, 123907y
by *Aloe arborescens*, crassulacean acid metabolism

FIGURE H. Chemical Abstracts, "General Subject Index": vol. 77, 1972. Types of entries: 1) Organisms, 2) Body parts and structures, 3) Processes, phenomena, and concepts.

catch--word phrases that eptomize the subject of the article or that aspect of the article being indexed. They are not necessarily part of the title. Because these phrases are taken from the text of the title, abstract or original article, all items on the same specific aspect of the compound are not always together. For example, seven different articles on the effect of DDT on bird reproduction are widely scattered under the Biologi--cal Studies heading.

4. The number following each entry is the volume number and abstract number.
5. All entries can be assumed to be re--search reports unless a capital letter, B (Book), P (Patent), or R (Review) precedes the abstract number.

The emphasis on compounds should not mislead you. There are many entries in the "General Sub-jects Index" which are useful to biologists. FIGURE H illustrates the major categories which you will find: 1) Organisms, 2) Body parts and structures (concrete entities of an organism), 3) Processes, concepts and other phenomena.

Conclusion

Chemical Abstracts is a complex and sometimes difficult tool to use. However, the high quality of its indexing and abstracting and the compre--hensiveness of the coverage make its use a must for any search which has a chemical aspect. When you use it consult with the reference librarian for help. Even the most advanced chemist can have trouble using it!!

APPENDIX VI

Zoological Record

In Chapter 6 we mentioned a number of indexes which are generally useful in biology. In addition to these there are several other more specialized indexes which are only useful for searches in certain areas of biology. Many of these indexes are listed in Appendix IV along with other reference sources in particular subjects. However, the *Zoological Record* is too important to relegate to just a listing. Therefore the following discussion explains briefly how and when to use it.

When?

Unfortunately most schools without graduate programs do not have *Z.R.* Nevertheless it is an important tool for locating literature about a specific organism group: genus, family, order. Anyone trying to do a comprehensive search of the biological literature for material on a particular animal group should consult the *Record* in addition to *Biological Abstracts, Bioresearch Index,* and other indexes. However coverage of a particular year's literature appears as much as three years later and therefore it is not useful for locating the latest information.

What is it?

Zoological Record is actually many indexes in one. It consists of 20 sections (FIGURE I) each covering a different animal group. The section for each group contains five parts (FIGURE J).

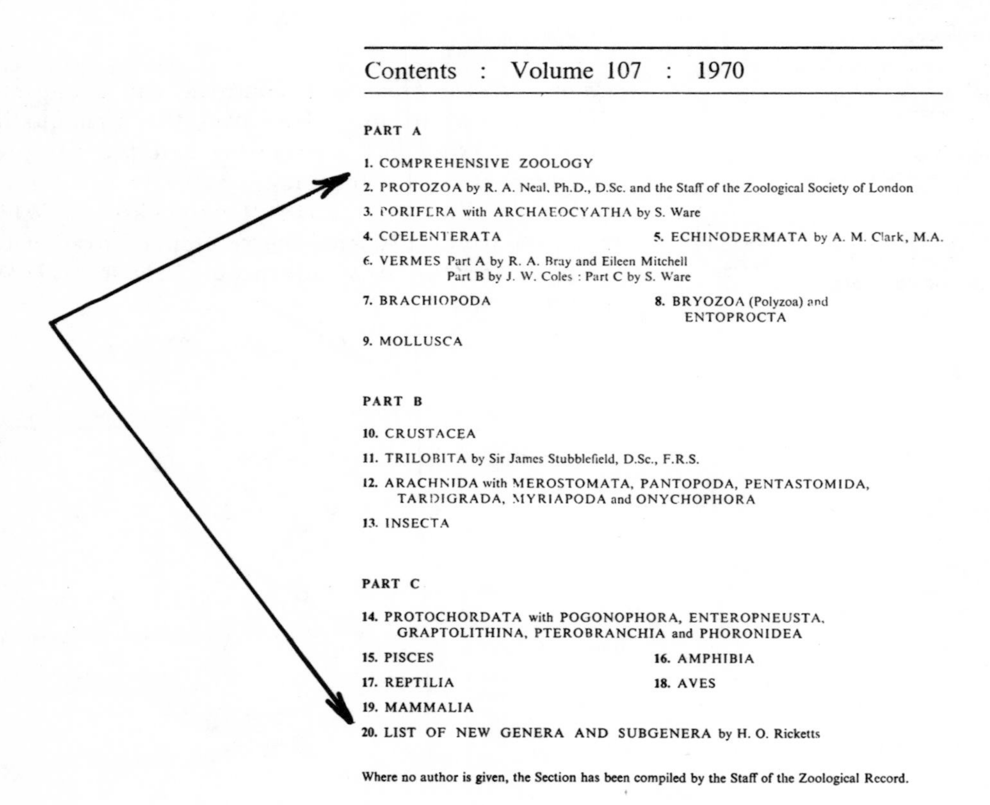

Contents : Volume 107 : 1970

PART A

1. COMPREHENSIVE ZOOLOGY
2. PROTOZOA by R. A. Neal, Ph.D., D.Sc. and the Staff of the Zoological Society of London
3. PORIFERA with ARCHAEOCYATHA by S. Ware
4. COELENTERATA
5. ECHINODERMATA by A. M. Clark, M.A.
6. VERMES Part A by R. A. Bray and Eileen Mitchell; Part B by J. W. Coles : Part C by S. Ware
7. BRACHIOPODA
8. BRYOZOA (Polyzoa) and ENTOPROCTA
9. MOLLUSCA

PART B

10. CRUSTACEA
11. TRILOBITA by Sir James Stubblefield, D.Sc., F.R.S.
12. ARACHNIDA with MEROSTOMATA, PANTOPODA, PENTASTOMIDA, TARDIGRADA, MYRIAPODA and ONYCHOPHORA
13. INSECTA

PART C

14. PROTOCHORDATA with POGONOPHORA, ENTEROPNEUSTA, GRAPTOLITHINA, PTEROBRANCHIA and PHORONIDEA
15. PISCES
16. AMPHIBIA
17. REPTILIA
18. AVES
19. MAMMALIA
20. LIST OF NEW GENERA AND SUBGENERA by H. O. Ricketts

Where no author is given, the Section has been compiled by the Staff of the Zoological Record.

FIGURE I. Zoological Record, List of Sections.

19. Mammalia : Contents

1. AUTHOR INDEX

2. SUBJECT INDEX
(see detailed index p. vii)

3. GEOGRAPHICAL INDEX

4. PALAEONTOLOGICAL INDEX

5. SYSTEMATIC INDEX
(see detailed index p. xi)

FIGURE J. Zoological Record, Section 19. Mammalia, "Contents".

The first part consists of an alphabetical listing of all citations by author, with a full bibliographic citation (FIGURE K). There are cross references from junior authors to the senior author. Following the "Author Index" there is a "Subject Index." This is a classified index that groups headings under broad subject categories (FIGURE L). Therefore headings like "Competition" or "Predation" will appear under the general heading "Ecology." (There is an alphabetical list of specific subjects which refers the users to the general category where it is listed.) In each subject category the articles are arranged by organism (using the scientific name). For each article the author's name is given (FIGURE L). The third part is a "Systematic Index" which lists articles by genus and species with genera listed under the Family; Families listed under the Suborder; and Sub-

Jordan, M. *see* Meyer, M.

Jordan, P. A. & Botkin, D. B. Biomass dynamics in a moose population.
Ecology **52**: 147–152, 2 figs, 1 tab.

Jordan, P. A. *see* Berwick, S. H.

Jörg, E. Ein Cranium von *Coelodonta antiquitatis* (Blumenb.) (Perissodactyla, Mamm.) aus plesitozänen Neckarkiesen von Mannheim-Käfertal.
Abh. hess. Landesamt. Bodenforsch. **60**: 83–88, 3 pls.

FIGURE K. Zoological Record, Section 19. Mammalia, vol. 108, 1971. "Author Index".

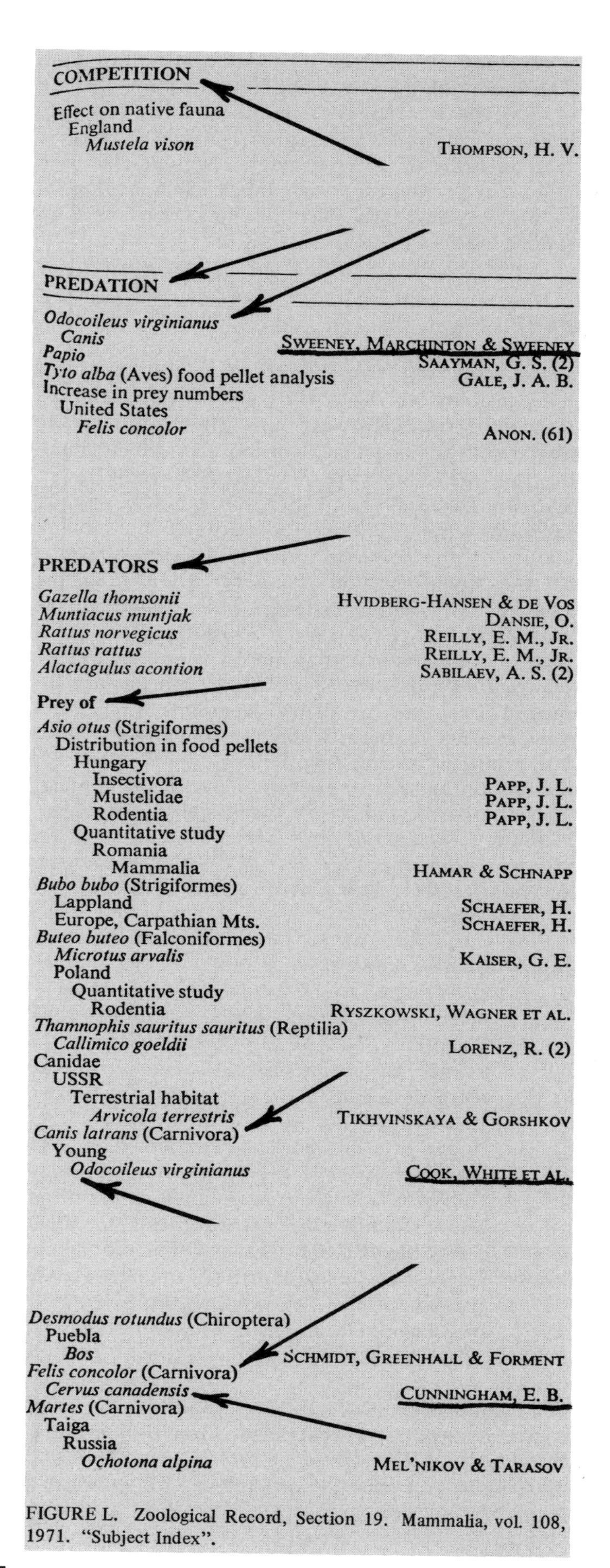

COMPETITION
Effect on native fauna
England
Mustela vison — THOMPSON, H. V.

PREDATION
Odocoileus virginianus
Canis — SWEENEY, MARCHINTON & SWEENEY
Papio — SAAYMAN, G. S. (2)
Tyto alba (Aves) food pellet analysis — GALE, J. A. B.
Increase in prey numbers
United States
Felis concolor — ANON. (61)

PREDATORS
Gazella thomsonii — HVIDBERG-HANSEN & DE VOS
Muntiacus muntjak — DANSIE, O.
Rattus norvegicus — REILLY, E. M., JR.
Rattus rattus — REILLY, E. M., JR.
Alactagulus acontion — SABILAEV, A. S. (2)

Prey of
Asio otus (Strigiformes)
Distribution in food pellets
Hungary
Insectivora — PAPP, J. L.
Mustelidae — PAPP, J. L.
Rodentia — PAPP, J. L.
Quantitative study
Romania
Mammalia — HAMAR & SCHNAPP
Bubo bubo (Strigiformes)
Lappland — SCHAEFER, H.
Europe, Carpathian Mts. — SCHAEFER, H.
Buteo buteo (Falconiformes)
Microtus arvalis — KAISER, G. E.
Poland
Quantitative study
Rodentia — RYSZKOWSKI, WAGNER ET AL.
Thamnophis sauritus sauritus (Reptilia)
Callimico goeldii — LORENZ, R. (2)
Canidae
USSR
Terrestrial habitat
Arvicola terrestris — TIKHVINSKAYA & GORSHKOV
Canis latrans (Carnivora)
Young
Odocoileus virginianus — COOK, WHITE ET AL.

Desmodus rotundus (Chiroptera)
Puebla
Bos — SCHMIDT, GREENHALL & FORMENT
Felis concolor (Carnivora)
Cervus canadensis — CUNNINGHAM, E. B.
Martes (Carnivora)
Taiga
Russia
Ochotona alpina — MEL'NIKOV & TARASOV

FIGURE L. Zoological Record, Section 19. Mammalia, vol. 108, 1971. "Subject Index".

orders under the Order (FIGURE M). For each article something about the aspect of the biology of the organism that is the main topic of the article is given, followed by the author's name. The other two portions of *Z.R.* provide a similar classified index for geographic areas, and a Paleonotological Index for geologic time periods which are in reverse chronological order.

How?

Since there are five indexes naturally there are five approaches. And while the "Author Index" is quite straight--forward the other indexes are not as self explanatory. It is important to remem-- ber that the indexes are in a classified arrangement. This has advantages and disadvantages. If you are thoroughly familiar with the status of the nomen– clature of your organism and/or the subject aspect of the organism you want to study then the arrangement is very helpful. The classified ar– rangement brings together articles on either related groups or topics and thus permits you, for exam-- ple, to find either the same kind of information on related organisms, or all the organisms which have been studied for a particular subject. It also avoids the problem of tracking down all synonyms, or related terms that represent a particular subject. This is not an index for the neophyte biologist. Without a grasp of the basic subject you are studying and the location of your organism systematics--wise the classified arrangements can be cumbersome.

Begin the use of the *Zoological Record* by following these steps:

1. Select the section which includes your organism (FIGURE I).
2. From the Table of Contents for that section select the approach you want to use (FIGURE J).
3. In the relevant parts of the index, choose the most specific categories that represent your topic. You must then scan every listing within that category and *all* sub– categories (FIGURES L and M).
4. When you have identified an article that appears useful, note the author's name and check it in the "Author Index" (FIGURE K).

The *Zoological Record* is important in doing complete studies of particular organism groups. It is more convenient than *Biological Abstracts,* particularly for the period before the establish-- ment of *B.A.*'s separate "Generic Index" in 1974. *Z.R.* began in 1865 and therefore it covers a larger body of literature than *B.A.* Finally, the classified approach permits easier access to material on related organisms, or subjects.

ORDER ARTIODACTYLA	
Key to families	
Ethiopian region, p. 7	ANSELL, W. F. H.
Museum exhibit	
Italy	
Genoa	MENCHINELLI, R.
Agriculture	
Husbandry of wild populations	
Ethiopian region	NASIMOVICH, A. A.
CERVIDAE	
Education and entertainment	
Recreational asset	GRANT, B.
Effect of man on natural populations	
Endangered species	WHITEHEAD, G. K.
Deer management	
United Kingdom	SMITH, P.
Artificial feeding	
Advantages and disadvantages	MACNALLY, L.
Hunting for sport	
France, Cerisy	
Population regulation	DUNCOMBE, F.
Australia	BENTLEY, A.
Alces alces andersoni	
Energy flow in ecosyst...	
Biomass dynamics	
Michigan	JORDAN & BOTKIN
Predator prey relations	
Michigan	JORDAN & BOTKIN
†*Alces latifrons*	
Stratigraphical indicator	
Pleistocene	
Moldavia, Tiraspol gravels	VERESHCHAGIN & DAVID
Cervus	
Herma...	TRATZ, E. P.
Cervus canadensis	
Conservation	
Scotland	STAINES, B.
Damage to agricultural produce	
England, Exmoor	LLOYD, E. R.
Animals as sport	
England, Exmoor	LLOYD, E. R.
Prey of	
Felis concolor	CUNNINGHAM, E. B.
Cervus canadensis nelsoni	
Prey of	
Ursus am... (Carnivora)	
Wyoming	BANMORE & STRADLEY
Herbivorous feeding	
Idaho Primitive area	
Ecology	WING, L. D.
Cervus duvauceli branderi	
Protection of endangered species	
Rehabilitation in Kanha National Park, Madhya Pradesh	
India	BINNEY, BOURLIÈRE ET AL.

FIGURE M. Zoological Record, Section 19. Mammalia, vol. 108, 1971. "Systematic Index."

Index of Titles

Note: This index includes only the *reference* sources treated in the nine chapters. It excludes non--reference books and periodicals and reference works cited only in the Appendices II, III, and IV.

Notes

Notes

Notes

Notes

Notes